PRAISE FOR THE WOR

BEATING T

"Enjoyable and instructive . . . [Dr. Ramsland] makes these forensic initiatives, and the inventors and investigators who nourished them, really come to life."

—Taph.com

THE DEVIL'S DOZEN

"Highlights [Ramsland's] command of history and forensic knowledge, fusing it with her interesting and approachable narrative voice . . . *The Devil's Dozen* is a collection of some of the most fascinating cases and instances of innovative investigation." —BioMedSearch.com

THE HUMAN PREDATOR

"If you're looking for the perfect gift for someone who's riveted to television shows like *CSI*, you won't find a better one . . . This book is unique in the field."

—*truTV Crime Library*

THE C.S.I. EFFECT

"A fascinating must read for *CSI* fans and anyone interested in criminal justice." —*Booklist*

continued . . .

THE UNKNOWN DARKNESS: PROFILING THE PREDATORS AMONG US

Coauthored with Gregg O. McCrary

"A must read for true crime fans. A beautifully written expert analysis of high-profile killers." —Ann Rule

"One of the most immensely readable and gripping accounts of serial murder I have ever read."

—Colin Wilson, author of *Serial Killers: A Study in the Psychology of Violence*

THE FORENSIC SCIENCE OF C.S.I.

"Fascinating . . . [A] must for anyone who wonders how the real crime solvers do it." —Michael Palmer

"With the mind of a true investigator, Ramsland demystifies the world of forensics with authentic and vivid detail." —John Douglas

PIERCING THE DARKNESS: UNDERCOVER WITH VAMPIRES IN AMERICA TODAY

"A riveting read, a model of engaged journalism."

—*Publishers Weekly*

Titles by Katherine Ramsland

THE FORENSIC PSYCHOLOGY OF CRIMINAL MINDS

THE DEVIL'S DOZEN

TRUE STORIES OF C.S.I.

BEATING THE DEVIL'S GAME

THE C.S.I. EFFECT

THE HUMAN PREDATOR:
A HISTORICAL CHRONICLE OF SERIAL MURDER
AND FORENSIC INVESTIGATION

INSIDE THE MINDS OF MASS MURDERERS: WHY THEY KILL

A VOICE FOR THE DEAD:
A FORENSIC INVESTIGATOR'S PURSUIT OF TRUTH IN THE GRAVE
(with James E. Starrs)

THE SCIENCE OF COLD CASE FILES

THE UNKNOWN DARKNESS:
PROFILING THE PREDATORS AMONG US
(with Gregg O. McCrary)

THE SCIENCE OF VAMPIRES

THE CRIMINAL MIND:
A WRITER'S GUIDE TO FORENSIC PSYCHOLOGY

THE FORENSIC SCIENCE OF C.S.I.

CEMETERY STORIES:
HAUNTED GRAVEYARDS, EMBALMING SECRETS,
AND THE LIFE OF A CORPSE AFTER DEATH

GHOST: INVESTIGATING THE OTHER SIDE

BLISS: WRITING TO FIND YOUR TRUE SELF

PIERCING THE DARKNESS:
UNDERCOVER WITH VAMPIRES IN AMERICA TODAY

DEAN KOONTZ: A WRITER'S BIOGRAPHY

PRISM OF THE NIGHT: A BIOGRAPHY OF ANNE RICE

THE WITCHES' COMPANION:
THE OFFICIAL GUIDE TO
ANNE RICE'S LIVES OF THE MAYFAIR WITCHES

THE VAMPIRE COMPANION:
THE OFFICIAL GUIDE TO
ANNE RICE'S THE VAMPIRE CHRONICLES

THE ANNE RICE READER

THE ART OF LEARNING: A SELF-HELP MANUAL FOR STUDENTS

ENGAGING THE IMMEDIATE:
APPLYING KIERKEGAARD'S INDIRECT COMMUNICATION
TO THE PRACTICE OF PSYCHOTHERAPY

Novels

THE BLOOD HUNTERS

THE HEAT SEEKERS

BEATING THE DEVIL'S GAME

A HISTORY OF FORENSIC SCIENCE AND CRIMINAL INVESTIGATION

Katherine Ramsland, Ph.D.

BERKLEY BOOKS, NEW YORK

THE BERKLEY PUBLISHING GROUP
Published by the Penguin Group
Penguin Group (USA) LLC
375 Hudson Street, New York, New York 10014

USA • Canada • UK • Ireland • Australia • New Zealand • India • South Africa • China

penguin.com

A Penguin Random House Company

BEATING THE DEVIL'S GAME

A Berkley Book / published by arrangement with the author

For information, address: The Berkley Publishing Group,
a division of Penguin Group (USA) LLC,
375 Hudson Street, New York, New York 10014.

ISBN: 978-0-425-27145-2

PUBLISHING HISTORY
Berkley hardcover edition / September 2007
Berkley trade paperback edition / June 2008
Berkley premium edition / February 2014

PRINTED IN THE UNITED STATES OF AMERICA

10 9 8 7 6 5 4 3 2 1

Cover art by Gary S. Chapman/GettyImages.

To my parents,
for their influence and support

CONTENTS

ACKNOWLEDGMENTS

This book is the result of years of interaction with many people, as well as opportunities to research and write about forensic science. I'll probably miss someone, but among those whom I wish to acknowledge are:

John Silbersack, my multitalented literary agent and close friend, who understands me, maintains our vision, and keeps the momentum going.

Marilyn Bardsley, my editor and friend at Crime Library, who got me started writing about forensic science.

Ruth Osborne, who listens to me describe what I'm writing, no matter how gruesome, and who proofread portions of the manuscript.

Dana DeVito, my "coroner sister," who helps me be even more gruesome.

John Timpane, who has encouraged me every step of the way and provided material that was difficult to find.

Gregg McCrary, who taught me so much about profiling and investigation.

Robert Hare, whose work I deeply respect, and who

provided me with great resources for understanding psychopaths.

James E. Starrs, who got me involved with the American Academy of Forensic Sciences, not to mention let me be on his exhumation teams.

Forensic scientists, with whom I've had many conversations: Traci Starrs, Karen Taylor, Jack Frost, Zack Lysek, Tom Crist, Henry Lee, Jack Levisky, Tim Palmbach, Michael Baden, Cyril Wecht, and Bill Bass.

Deborah Brown, whose friendship steered me toward the Nutshell Studies.

Kim Lionetti, who made the early push for me in publishing forensics books.

My mother, Barbara, who never discouraged me from my "dark interests" as I grew up, and my father, Henry, who always supported my education.

Karen Pepper, who provided research support and proofreading.

Karen Walton, who always reads my books, even when they scare her.

Ginjer Buchanan, my enthusiastic editor at Berkley, for her encouragement and interest in all that I'm doing.

INTRODUCTION

Early in the morning of June 13, 1994, someone attacked and killed Nicole Brown Simpson and Ronald Goldman outside Simpson's home in Brentwood, California. Both had been slashed to death with a knife, and from the extreme brutality of the attack, Nicole appeared to have been the target. That threw suspicion on her estranged ex-husband, former celebrity athlete O. J. Simpson. His arrest and trial brought to the public's awareness the nature of evidence handling and scientific analysis in a criminal investigation, during an age when a number of sophisticated procedures were being utilized. Spots of blood turned up in Simpson's vehicle and home, and he had a cut hand supposedly from the same night as the incident, but he protested that he was not guilty. Still, he did attempt to flee with a bundle of money and he penned a suicide note. He also had opportunity to commit the double homicide, as well as a history of threats and domestic violence involving Nicole.

Simpson's murder trial was televised, offering viewers months of proceedings that featured crime scene specialists, crime lab personnel, pathologists, investigators, DNA analysts, blood spatter pattern analysts, and other experts who were called on to reconstruct the crime. In the end Simpson was acquitted, while investigators, scientists, and lab personnel were left to ponder their costly and embarrassing mistakes. Not only had the public learned many details about the investigative process but they'd also seen its foibles and limitations. Nevertheless, widespread interest in forensic investigation had been sparked, leading the way for the field of forensic science to garner more attention, including funding from Congress. We'll return to the Simpson investigation later in the book, since it's part of the fascinating history of just how science and the legal process merged. Our focus will be on those people—criminals, investigators, and scientists—whose lives and activities affected one another toward the goal of improving the pursuit of justice and our understanding of why some people criminally harm others.

The priority in forensic investigation is to make an identification, whether it be of a victim, an offender, a fraudulent document, or a weapon. Before scientific analysis was established, people were identified and crimes were solved mostly through logic, snitches, and confessions. For centuries in fact, among the best methods for getting a confession was torture. If it made sense that a specific individual had committed a crime and he or she wasn't talking, there were ways to persuade him. Suspects might be stretched, roasted, dunked in water, whipped, branded, or subjected to having a body part chopped off, and it's likely that to stop the pain or prevent disfiguration, many an innocent person blurted out whatever was required. Punishments ranged from banishment from a community to death by grisly means.

Yet not all convicted criminals were detained for life or executed, and thus society needed a way to identify repeat offenders. Branding was effective, as were certain types of bodily mutilation, but some cultures devised a means of describing criminals so as to recognize them if they returned before governing bodies. Even so, most systematic forms of identification and categorization did not develop until the nineteenth century.

The story of forensic science unites the narratives of both science and law. It's important to see first how each developed to fully appreciate how their merging proved productive for both. Science emerged as both a perspective on human experience and an objective measure of human reality. Guided by the fourteenth-century notion of Occam's razor, which encouraged devising the simplest explanation for phenomena, persistent men who found more truth in the scientific method than in religion, myth, or superstition, directed the evolution of science and law. They took risks to have their experiments and theories viewed seriously and their careful work often yielded impressive results. They were assisted by the empirical philosophers of the seventeenth and eighteenth centuries who shifted ideas away from esoteric notions about reality and toward experience and observation. Yet just as science did not achieve recognition without a struggle, the same is true of its forensic application.

In its earliest days, physicians, chemists, and other scientists attempted to develop and reform legal proceedings, which eventually inspired new investigative techniques and methods for dealing with offenders. While court systems initially resisted scientific procedures as a hindrance to logical analysis (especially when these approaches diverged), the impressive resolution of sensational cases with the help of science slowly persuaded the judiciary that scientific methodology was beneficial, then essential, to achieving justice.

Forensic science is the application of scientific perspectives and techniques to the legal process, including investigations and courtroom protocol. While it's lately become an umbrella term covering everything from the investigation of historic events that have no forensic significance to disciplines of skill that involve little to no science, it's most technically accurate to say that forensic science is the use of scientific data and procedures specifically for the legal system. This discipline relies on the same values and methodology used by other members of the scientific community. There's a rigorous procedure involved, including controlled conditions, reliable data collection, and the attempt to disprove hypotheses. A hypothesis must be testable, which means it must be capable of being proven wrong. In setting up a hypothesis, the scientists can envision other possible interpretations of the data, such that they can determine whether one hypothesis or another best applies. Too often investigators develop a hypothesis and then seek data to support it. They will likely find what they're looking for, but in doing so may twist the facts, or neglect or ignore others that could contribute to a different outcome.

Scientists are interested only in the outcome the facts allow, not in the outcome they may desire, because their work must be performed in a way that others can replicate: with objective methods and reasoning that other forensic scientists—their peers—would support.

"Observation" was the watchword of the early forensic pioneers. They understood the value of observing a crime scene as a whole and then identifying something in that context that should be carefully analyzed. It could be gunshot residue on someone's coat, a piece of bomb shrapnel, a strand of hair, or a suspicious stain on a carpet. It might be how a car was driven against a tree or blood spattered onto a wall. It might even be behavioral clues, such as the fact that three similar crimes all occurred in the same neighborhood between midnight and two A.M. The scientists focused on the subject of study and pondered its implications before they determined the appropriate method. A proven scientific procedure evolved for each type of evidence, whether measuring bitemark impressions, analyzing the DNA in saliva, or finding poison in human tissue, and the men who introduced and refined the best procedures contributed toward our betterment. (Until recently, this area has been the province of male scientists, so I'll generally use the male pronoun.)

THE EMERGENCE OF SCIENCE

For the philosopher Heraclitus in ancient Greece, the one key principle to explain everything in our world lay in the notion of change, while to Parmenides the principle was permanence. Yet others believed that reality was based in numbers and mathematics, and a few noted thinkers viewed it as mental or spiritual rather than physical. Still, all of this was largely theoretical. It was Aristotle who began to organize and systematize concepts about the physical world, and his approach long defined scientific methodology. The spiritual world was assigned to the realm of religion, and science looked to those objects and situations that could be observed, measured, and categorized.

Yet even the solid and secure evolved into something more variable. Throughout the history of science, ideological foundations that defined our notions of reality have been shattered and rebuilt, as emerging anomalous facts revolutionized even the most entrenched perspectives. Many theories once believed fundamental have been replaced with others that better explain the body of facts and observations. Then with new knowledge that reveals the limitations of even *those* ideas, the theories give way.

The bulk of forensic science developed during the era of Newtonian physics in the nineteenth century, with key influences from Charles Darwin, physical anthropology, medicine, and even early psychiatry. The innovations took hold around mid-century and picked up speed over the next hundred years. Instead of attempting to include every possible development in every area of forensic science or technology in this volume, I've selected what I consider the most illuminating and interesting tales, setting them out as stories to give the chronology momentum. I was especially impressed with the vision of some key participants.

To explain and defeat the problem of crime, the early forensic scientists and investigators tended to focus on the criminal, with special study devoted to the criminally insane. They examined the criminals' physical, moral, and psychological constitution, as well as their hereditary and environmental influences. In general, the experts viewed criminals as set apart in some manner from normal society, whether it was how they looked or how they thought, and certain thinkers believed that if the cause of their deviant behavior could only be clarified as the manifestation of a "type" of person, crime could be erased and society thereby ennobled. Thus, a number of scientists set about to train themselves to closely observe incarcerated criminals so as to learn how to spot them "in the wild." This expertise alone, although there was more logic than science, gave them the status of a "scientist" and allowed them to speak with authority in their various societies, in treatises, professional journals, and in court. Some of these authorities explained criminality as a moral failure, others as a biopsychological defect, and still others as the effect of poor social conditions or the result of inheriting family corruption.

Nineteenth-century practices in forensic science derived from ideologies that proposed perfect societies free of crime, and they believed that specific people who were responsible for illegal deeds (mostly foreigners or people from the lower classes) could be singled out and dissected so as to decrease recidivism and prevent future criminality in others. They proposed that the state support experiments that, in the long run, would benefit society, paying back manifold any expenses incurred. Since they needed a norm against which to measure success, they devised the "rational man," who used his wits to exercise control over his behavior and strive to better himself. As a result, those men who offered informed opinions from scientific disciplines to achieve diminished offenses and greater moral self-improvement in the greatest number of people became increasingly important in the courtroom. That they might make a mess of things in the process did not stop the courts from increasing their participation and at times even compelling it. Once science got a foot through the courtroom door, there was no getting it out. The only thing self-respecting scientists could do was to keep improving their contributions and to police their colleagues.

Thus, as science assisted justice, justice helped to polish this form of applied science. Largely, that's the story of forensic science. But it's also the tale of visionaries who persisted in setting forth standards within the judicial arena. From the earliest autopsy results that figured in a criminal case to the latest technology for DNA analysis, forensic science has come a long way.

ONE

FORENSIC FIRSTS

PRIMITIVE PRACTICES

One of the early attempts to use measured evidence in a court of law occured during the Salem witch trials of 1692. In Salem, Massachusetts, Elizabeth Parris, nine, and Abigail Williams, eleven, had begun to act strangely and they soon accused neighbors and relatives of using witchcraft to assault them. A black slave, Tituba, was arrested and she admitted there was a conspiracy of witches in the area. Other children joined the first two in their convulsions, seizures, and trancelike behavior, and terror over the invasive presence of satanic forces spread throughout the town.

More people were arrested based on "spectral evidence" and forced confessions and were sentenced to death, among them the Reverend George Burroughs. He allegedly bit his victims while in spectral form, leaving a visible mark. To ensure that it was Burroughs and no other who left this impression, his mouth was forced open in court to compare against bite marks on the girls and against the teeth of other people present. The seven judges on the Court of Oyer and Terminer were satisfied that Burroughs could be identified by his teeth as the man who inflicted the torture via satanic means, so he was convicted and hanged with four others. In all, twenty people lost their lives before the governor finally prohibited the court from relying on intangible evidence.

Crime has been part of human society since Cain slew Abel, and identifying perpetrators has challenged the greatest minds to develop reliable techniques for investigation and prosecution. Some approaches, such as torture and mutilation, have contributed little to the advancement of justice, but others have inspired clever and unique methods of apprehension. For centuries, physicians led the way. During Imhotep's reign in Egypt in 2980 B.C., medical experts performed autopsies for illnesses and wounds, and in 535 B.C., the Greek physician Alcmaeon of Croton studied human cadavers, although the notion of science was not yet formalized.

Several ancient societies recognized the individualizing value of the ridged patterns of fingerprints. The earliest datable prints were left in Egypt about four thousand years ago, while the Chinese used prints to authenticate someone's craft. In the third century B.C., fingerprints or thumbprints showed up on official documents for business or court dealings, and in Rome, the attorney Quintillion relied on bloody fingerprints to prove that someone had tried to frame a blind man for his mother's murder. Yet no written record makes it clear that anyone regarded fingerprints as a way to individualize and identify people. It would be a long time before someone realized that the uniqueness of fingerprints had reliable forensic implications.

Shortly after Socrates drank a fatal dose of hemlock in 399 B.C. and Hippocrates discussed lethal wounds and prescribed arsenic remedies, the earliest murders by poisoning were documented, introducing the first primitive efforts in what would become the field of toxicology. The Alexandrian physicians Eraistratus and Herophilus set precedents for later medicolegal investigations of wounds, poison, and other mechanisms of death. In 82 B.C., the Roman Republic enacted the first law against poisoning and two decades later prosecuted a murder suspect, Cluentius, with the orator Cicero chosen to defend him. Cluentius supposedly had poisoned his stepfather, Oppianicus. Rather than defend Cluentius on the grounds of innocence, Cicero listed the victim's many known offenses, including multiple murders. Cicero used a clever bit of logic to leave the impression that the death of Oppianicus was just, in a perverse sort of way. Despite incriminating circumstances, Cluentius was not convicted.

The Republic and Empire of Rome endured from roughly 753 B.C., with the founding of Rome, until its collapse in 1453 A.D. Military garrisons and newly built roads had taken the practices and philosophies of Rome to distant lands, and other cultures soon acquired its law and utilitarian culture. In 54 B.C., Julius Caesar became ruler of the Roman world, and he extended Roman rule west to the Atlantic Ocean and as far north as England. His planned reforms upset the aristocracy, so a conspiracy of senators gathered on the Ides of March in 44 B.C. to assassinate him. Members of the conspiracy stabbed Caesar twenty-three times and he fell to the floor and died. His body was carried to a place where the physician Antistius could perform an autopsy. He determined that the single stab wound that had entered the heart had been the fatal one, announcing this before the forum, the arena that is now the basis for the word *forensic.*

After Caesar's death, a civil war arose among the various political factions, which ended with the establishment of the Second Triumvirate, and three men initially shared control over the Empire. Caesar's great-nephew, Caesar Augustus, emerged the victor. During this era, medicolegal medicine made few strides, apart from the Emperor Justinian's recognition that physicians could be "expert" witnesses.

The five centuries between 500 and 1000 A.D. are referred to as the "Dark Ages." Many of civilization's gains receded as trade and intellectual life declined. Christian scholars preserved the Church's teachings, along with the Latin language, but they guarded the skills and knowledge among an elite few, the self-endowed arbiters of God's truth.

During the Middle Ages in Western lands, the only real methodology practiced as a "science" was alchemy, which originated in ancient Egypt and integrated nature religions with primitive psychology, Greek Rationalism, astrology, and Egyptian chemistry into a mystical approach to "soul refinement." Egyptians believed that certain rituals ensured the right kind of life after death, which entailed preserving the corpse so the body could be transformed. Mummification supposedly bathed the corpse in "god-liquid," which had to be done during the right alignment of constellations—the moment favored by magic. Alchemists adopted this idea and adapted it into a Christianized notion that the soul could be ritualistically purified.

Relying on the opportune "inner moment" and specific chemical formulas, alchemists strove to produce a "Philosopher's Stone," which supposedly connected mind and matter to transform base metals into gold and opened up the soul to achieve unity with the infinite. An alchemist might busy himself crushing, heating, and dissolving substances in a laboratory while making careful records of the changes in both the physical substance and in himself. While "science" transformed via physical components, alchemy traveled a parallel inward route.

The first phase was the *nigredo*, or the black phase, wherein the alchemist surrendered to "dark" aspects of himself in order to learn and develop better awareness. In conjunction with this, he might break down complex physical substances into their most primitive and base forms. During the *albedo* (white) phase, the alchemist strove to purify himself and the physical materials for the yellow phase, the *citrinitas,* which helped make a transition to the red phase, the *rubedo.* Just as the sun rises in redness, so came enlightenment, rebirth, and the emergence of the Philosopher's Stone, the ultimate transformative substance that would perfect the soul and provide unlimited wealth. While no one claimed to have achieved this "Great Work," scholars who study this period believe that alchemical experiments with physical substances became a basis for some areas of chemistry. But before toxicology developed, death investigation itself became a more formalized practice, in both the West and the East.

DEATH INVESTIGATORS

Prior to the *Magna Carta*, England's *Charts of Privileges* listed the office that would evolve into coroner in 925 A.D. People taking up these duties were appointed throughout the country, and in September 1194, the judicial circuit in medieval England, under the rule of King Richard and in attunement with legal reforms, officially recorded in the *Articles of Eyre* that "crowners" would be elected to deal with certain judicial functions. They were the *custodes placitorum coronæ*, or the keepers of the king's pleas. Those who held the office, soon to be called coroners, collected taxes, but they also summoned inquest juries for people who were seriously wounded or who had died from "misadventure." Since these officials were there to protect the king's interests, they could confiscate animals or objects implicated in accidental deaths and take over goods found in accidents or wrecks, although they could not themselves render verdicts.

An inquest involved the gathering of seven to eleven

"good" men, who voted on a verdict, although the coroner's evidence was a strong factor in their decision. While coroners weren't physicians, they acted as unofficial medical examiners and could also issue orders for arrest.

One's manner of death had implications for taxes, because in certain types of deaths the king confiscated the property. This led to potential corruption by coroners: For a fee, the coroner might "define" the cause of death as something that would benefit the relatives. Eventually the office, which remained political, evolved into that of a death investigator, and by the thirteenth century, coroners examined all dead bodies to determine the nature of wounds or diseases and a person's manner of death. In the interests of community health or justice, they summoned physicians to perform autopsies. This relationship could produce tension between coroners and physicians, because the latter were often not paid a fair price (or any) for their services.

Since there was little by way of systematic medical procedures, coroners generally relied on what the circumstances indicated. Thus, people caught in the wrong place at the wrong time, or erroneously identified by eyewitnesses, could be imprisoned for acts they did not commit. It seems also to be the case that few doctors offered neutral testimony whenever they were involved; instead, they took sides in the adversarial arena. Indeed, there was no official body of practitioners who called for objective methods or practices.

In some areas, the law required that the person who found a body was to alert four people nearest to the body's location. They would let an official know, who would then notify the coroner. That person would view the body and assemble a coroner's jury, who would decide on whether death had been the result of natural causes, an accident, or otherwise. Reports were drawn up to record where the body had been found, whether the person had been killed on that spot, how the person had died, where he or she was last seen alive, and the location of the wounds. Medical experts might assist the judiciary in arriving at a decision, although that practice had already begun in Italy during the thirteenth century when a specialty in legal medicine was offered at the University of Bologna and Pope Gregory IX had recognized the contribution of physicians in the legal process. (Pope Innocent III, too, had appointed doctors to serve in the courts.) Bologna instituted an official medicolegal officer and people in these positions attested to the seriousness of wounds, especially those that did not cause immediate death. By the end of the century, the Iberian Peninsula followed this practice, and certain physicians developed greater expertise in medical-legal matters.

WASHING WRONGS AWAY

When the death investigator in thirteenth-century China came to examine the corpse of a person who had recently died from apparent homicide, he would make every effort to ascertain that the death was indeed unnatural. Among his questions might be queries about this person's known enemies but also anyone who might gain from making those individuals suspects in such a crime. Experienced investigators knew that a situation was not necessarily what it seemed and even natural deaths might be exploited to gain revenge against someone.

While ancient Rome grounded the European practice of law, China made the earliest strides in murder and manslaughter investigations, recording facts about the process of death and inventing mechanisms for crime detection. In 1247 A.D., Sung Tz'u, a Chinese lawyer, offered advice in one of the oldest extant works of forensic techniques, *Hsi yüan chi lu* (*The Washing Away of Unjust Imputations*), basing his ideas for solving cases and calculating decomposition rates on strict observation and logic. "Among criminal matters," the book opens, "none is more serious than capital cases; in capital cases nothing is given more weight than the initially collected facts; as to these initially collected facts nothing is more crucial than the holding of inquests."

Sung Tz'u was a national university Ph.D. graduate and a Judicial Intendant, and he derived his ideas and descriptions from death investigations dating from as early as 907 A.D. By 995, a procedure had been decreed that in the event of a death, an inquest—*Chien-yen*—was to be carried out, and if it appeared complicated (murder, abuse of prisoners, disease), then a high official would do a *fu-chien*, or further inquest. This involved an examination of the deceased, and until the inquest was complete the family would not receive the remains. If the official made a mistake in judgment regarding the cause or manner of death, he could be sanctioned with a prison sentence.

In this handbook on autopsies, the author described how causes of death such as drowning and strangulation alter the appearance of a body, documenting wounds and discolorations. Obvious suspects perpetrated most crimes, but to address the more mysterious incidents, systematic observations were offered to assist death investigators in making their judgments. Thus, they learned to distinguish deliberate deaths from suicide or accidental deaths, and to spot staged homicides.

A handful of civil servants oversaw the maintenance of order and the negotiation of legal disputes, and local authorities were assigned to a three-rung hierarchy. On one rung was a *hsien-wei*, or sheriff, for enforcing the law. Often they had no training, so handbooks such as the *Hsi yüan chi lu* helped them to learn about the elements of a death scene and how to locate and interrogate people who might have information. The process involved questioning the original informant (the finder of the body or witness to a homicide or suicide), other witnesses, relatives of the deceased, and suspects. Male victims received an examination from men, and women from midwives. If a mortally wounded person was in the process of dying, then a *pao-ku*, or "death limit," would be set that would stipulate, depending on how long it took the victim to die, whether the attacker would be charged with murder or assault.

Examinations were public, conducted with a *wu-tso*, or coroner's assistant, who also performed the burial, and everything was recorded in complex documents, all signed by witnesses. Standard drawings of body forms, similar to those used today, were used by examining officials to illustrate the wounds.

The inquest procedure was similar to a trial, with suspects openly confronted or placed under conditions of stress that might provoke a confession. The body was often present as well, as were the victim's relatives. Witnesses might then present information or evidence to assist the process. The guidebook also offered ethical guidelines, such as the need for officials to take care not to stay in the homes of people associated with the victim, lest their judgment be compromised.

Where evidence was minimal, Sung Tz'u provided scenarios to help officials learn to think their way through a potential crime. Contemporary forensic authors like to retell one tale in particular: A group of men was laboring in a ditch when one worker killed another. The victim was later found by the roadside, smitten with about ten sickle wounds. He still had his personal effects, so it seemed likely that the perpetrator knew him rather than it being a thief attacking a random stranger. The victim's wife indicated that another man had asked to borrow some money and was angry when refused, so a suspect profile was developed. However, during the inquest no one would confess, and further questions turned up nothing more, so the magistrate ordered the men of the village to gather together before him with their sickles. Some eighty men arrived and laid down the implements. In time, the flies came, attracted to blood and specks of flesh, and the sickle on which they alighted revealed the killer. He was the same man who had wanted to borrow money. Confronted with the evidence against him, he "knocked his head on the ground and confessed."

Sung Tz'u also relied on creative intuition to come up with a solution. In another scenario, he mentions that A wants to rob B, so when they cross water, A drowns B, leaving no marks. How can such a case be examined, he asks? "First, look to see if the body is emaciated, if the thumbnails and fingernails are black in color, whether there is sand or mud under the nails or in the nostrils [and] whether the chest is red . . ." If these conditions hold, then B was inferior in strength and was held under the water. To solve the crime, the examiner needed to learn about A's motives and to gather B's belongings as evidence, but the body would yield evidence as well to support the finding of murder, as long as one knew what to look for.

There were special instructions in the handbook for dealing with the killing of children and fetuses, as well as for handling bones, which was a particularly difficult task. Some bodies were admittedly too decomposed to serve as evidence (still true today). Sung Tz'u documented the differences among victims of poisoning, hanging, self-immolation, and other conditions. Thorough, meticulous, and clinical, the manual served death investigators for another seven centuries, with new information added as it was uncovered.

The Chinese also employed a primitive lie test in the form of requiring suspects in a crime to chew rice as they were questioned about an incident. Then they were instructed to spit it out. The assumption was that the offender would be unable to do so because stress would have caused the saliva in his mouth to dry up.

Since China was a closed society, these methods failed to have any impact on Western forensic practices, but eventually advances were made in Europe as well, albeit a few centuries later.

EXPLORING THE BODY

The tight union of Catholicism and politics across Europe hindered physicians from working on cadavers—the "sacred" body—so they could improve on ancient medical findings. Some dissected illegally, while others became inventive within the legal limits. After the Black Death decimated European society, killing some twenty-five million people during the fourteenth century, social classes reorganized, inspiring innovations not seen in several centuries. As a result, a new approach to practical life eroded traditional doctrines, and religion began to lose ground. In *City of God,* St. Augustine urged good Christians to beware of intellectual argument and scientific practices as a danger to their soul. Apparently he thought that using the mind was directly opposed to true faith, and while that attitude dominated Europe for centuries, philosophers like Roger Bacon urged people to rely on their reasoning powers. Sometimes the church condemned such heretics and even executed them, but eventually there were too many voices for the religious faction to halt the momentum.

New banking systems and increased trade encouraged a new optimism across Europe, along with exploration and discovery. During the Renaissance, which started in Italy around the middle of the fourteenth century, educated men like Leonardo da Vinci studied the humanities, turned to experimentation, and explored the biological structure of the human body to learn more about its functions. The Catholic Church continued to resist this trend, and while some scientists capitulated and denounced their own ideas, the plethora of discoveries about geology, astronomy, chemistry, and biology eventually reached critical mass. When Johann Gutenberg perfected movable type in 1453, he changed modes of communication and helped others to educate the masses. A new respect for the scientific method, based in theoretical mathematics, inspired discoveries such as the behavior of gravity and inventions such as telescopes, which assisted Galileo to confirm that Earth and other planets revolved around the sun.

Philosophers challenged the Church with humanistic ideas, while Luther birthed the Protestant revolt. Scientific progress soon became a fashion. Seafaring ventures brought diverse cultures and whole continents in contact with one another and Europeans settled other lands. Industrialism had its foot in the door of heretofore agrarian societies, gradually easing daily life with modernizing devices.

During this time in Europe, autopsies were performed

for forensic purposes and the early anatomists who dissected corpses noticed how conditions such as rigor mortis (the state of muscle rigidity) or algor mortis (the body's cooling temperature) worked, and to the list they also added the progression of coloration changes in a decomposing corpse, known as livor mortis, or lividity. A few physicians published early medical works about this subject, along with treatises more specific to investigation, such as the 1452 publication on gunshot wounds by Hieronymous Brunschwygk and the 1507 Bavarian text about the participation of physicians in legal cases, *Constitutio Bambergenis Criminalis*.

In France, Charles V decreed in the Carolingian Code of 1533 that courts utilize evidence from autopsies in specific types of cases, such as infanticide, homicide, and apparent poisoning. The Code authorized funds for medicolegal training, paid practitioners to apply medical science to matters of law, and urged them to use their knowledge to improve public health. Some who offered testimony also reported their cases to their colleagues in the medical community, and in 1560, the first scientific society was organized in Italy. The Netherlands, too, got into the act, by building an anatomy theater at the University of Leiden for public teaching via autopsy.

One of the more "scientific" practitioners of this era was Ambroise Paré, who toward the end of the sixteenth century acted as surgeon and medical advisor to King Charles IX of France. He studied injuries from firearms and had an interest in poisons, a popular means for disposing of someone. Yet he also kept a skeptical stance and was keen to debunk popular superstitions. In one situation, he decided to prove that a certain "cure" based on gallbladder stones cultivated from animals was a waste of the king's money. The palace cook, imprisoned for theft, was selected as the guinea pig. Paré fed the man some poison and then gave him the fake cure. But, as expected, the cure did not work and the cook died in agony by the end of the day. Still, despite the lack of ethics shown, it was a primitive demonstration in science.

By century's end, Battista Codronchi offered *De Morbis Veneficiis,* a study of poisoning deaths, while seven years later, Fortunato Fedele published *De Relationibus Medicorum,* and in Rome, Paolo Zacchia addressed these matters in the *Quaestiones Medico-Legales.* Frenchman Francois Demelle published the first study on handwriting analysis in 1609, while in 1628, William Harvey explained blood circulation, indicating for the first time that blood stayed in the body rather than being used up and renewed as was formerly believed. Toward the middle of that century Germany's University of Leipzig offered a course in forensic medicine, and around that same time, Italian biologist Marcello Malpighi described pattern ridges of fingerprints, although the discovery would not be forensically significant for two more centuries.

Also in the seventeenth century, French philosopher René Descartes divided all reality quite distinctly into two substances, mind and matter. He determined that the material world operated like a machine, and this mechanistic view soon became the dominant scientific perspective. Descartes used analytical geometry to draw pictures of the way time and distance worked together, and shortly thereafter Isaac Newton made this the foundation for formulating his ideas, among them the laws of motion. Newtonian mechanics defined scientific reality throughout the nineteenth century and into the early twentieth century. Thus, matter was viewed as essentially passive and all events were the result of a definite cause. Predictions about the physical world could be offered with confidence because the laws of the universe were thought invariable, and those ideas laid the basis for technology. Through accurate measurements, the physical world could be controlled and predicted and science was based in rationality, objectivity, and exactness. Rational knowledge derived from how people experienced and thought about objects in the environment through comparison, measurement, and classification. Intellectual distinctions and the assignment of incompatible opposites made the logic of science possible.

SEEING BETTER

Most significant at this time was the work of Anton Van Leeuwenhoek, who invented the world's first powerful precision microscope. The ancient Romans had used magnifying glasses, and Italian nobleman Salvino D'Armate, who also made concave lenses, donned the first wearable eyeglasses in 1284, but it would be three centuries before true microscopy was born. In 1590 in Middelburg, Holland, spectacle makers Zaccharias Janssen and his son, Hans, devised a combination of lenses by putting two different lenses at either end of a long tube to produce an even larger image via the first telescope. However, the image came through upside down, so another inventive mind, Johan Lippershey, made one lens concave to right the magnified image. Other glass manufacturers refined magnification through tiny lenses, which produced the name *microscope.* Galileo had improved their utility with a focusing device and more finely ground lenses to produce even higher magnification power.

Some time later Anton van Leeuwenhoek became the real father of microscopy (although Robert Hooke developed a weak form of the compound microscope five years before him). Around 1670, as part of a hobby, Leeuwenhoek devised curvatures that increased the magnification power of a lens and made numerous biological discoveries. In fact, he noted that a shaft of hair has a number of colors and scales. The stereoscopic microscope added a double eyepiece and prisms, which provided three-dimensional images. It wasn't long before Leeuwenhoek used this device to describe the structure of red blood cells.

Another man who utilized experimentation to look beyond a "fact" long taken for granted was Italian physician Francesco Redi. For a long time in Western countries, people believed that decaying corpses spontaneously produced flies, maggots, and beetles, because these insects were often seen on those that lay out in the open. Redi decided to make some observations of his own. He noticed that wrapped meat had fewer maggots than meat exposed to the air, so he placed fresh meat in three separate jars. One he left open, the other he covered with gauze, and the third he closed with a tight cap. When he found that the capped meat had no insect activity and the exposed meat had quite a lot, he realized that it was attracting flies that then laid eggs, which hatched into maggots. He did this experiment several times and was able to state that meat did not spontaneously produce maggots. Other people took up the study of the relationship between cadavers and insects, and in 1734 Rene de Reaumur published *A History of Insects,* an important milestone in forensic entomology.

EARLY CASES

Until the end of the eighteenth century in England, the accused represented themselves, and they often did not have the wherewithal to do so. Yet for a long time, there were no investigative methods, either, so the courts relied largely on a combination of eyewitness testimony and circumstantial logic.

Criminal investigation methods had emerged in England after Thomas de Veil was appointed magistrate for Westminster, and while he was among those who could be bribed, especially with the favors of a pretty woman, he also single-handedly brought down one of London's largest gang of thieves. He took up an office in Bow Street in 1739 and started to work on cases as an analyst, developing a keen instinct for figuring out criminal behavior and gaining renown in the process.

When a man disappeared in 1741, de Veil suspected the man's servant, James Hall, who told a less than convincing story about his master's whereabouts. De Veil pressured him and sent constables to search the missing man's place of business, Clement's Inn. They found the corpse, dumped head-first, in an outside privy. Hall broke down and admitted that he had killed the man, drained him of blood, and carried the body to the privy. Afterward, de Veil's reputation as a crime analyst spread and he was soon consulted on tricky crimes outside his jurisdiction that proved resistant to quick solutions.

After de Veil died, Sirs Henry and John Fielding, successive magistrates at the court at Bow Street, established the first constables in London in 1749. Henry Fielding had been a political playwright and novelist, but after he failed in the theater and his novels did poorly, he accepted a position as a justice of the peace. Unlike many of his predecessors, he truly did wish to eliminate crime from London and he set about to do that to the best of his ability. However, it proved to be a formidable task, since so many thieves and forgers had found their occupation an easy means to make money, with little consequence.

Henry Fielding's job was to track down known criminals and deliver writs of arrest. From a group of community-minded constables, he organized the "thief-takers" as a form of bounty hunter. Victims of crimes would report to Fielding and he'd send out a thief-taker to track down the culprit. For the first time, local robbers were identified, arrested, and sent to prison in an organized and efficient manner. Because the thief-takers were so quick to respond, they came to be known as the Bow Street Runners. The number of robberies in the area diminished. Fielding then sent armed constables against highwaymen and the results were similar. Committing crimes of this nature became riskier, even after Henry died in 1754 and his blind brother, John, took over. Collectively, their success was such that the office was expanded and proposals were made for routine patrols around the city. Nevertheless, there were no organized bodies for investigation, just the occasional brilliant thinker. Each new case proved the need for more systematic study of criminals.

This was evident to those who investigated murders by poison as well. An incident occurred in England in 1751, after a smitten Mary Blandy agreed to marry Captain William Cranstoun. She introduced him to her father as a man of wealth and position equal to his, and perhaps she herself did not then realize that he was already married, had no money, and was nothing short of a scoundrel hoping to enrich himself. He asked his wife to issue a statement that they were not married, and she initially agreed, but when Cranstoun started divorce proceedings, she balked. Cranstoun's dishonest scheme reached the ears of Mr. Blandy. He demanded that the lovebirds split up, but Mary was more inclined to carry on with the captain in secret, and to listen to *his* advice.

He urged Mary to place a white powder into her father's tea, but when it failed to dispatch him, Cranstoun told her to keep administering it in small doses. Blandy grew ill with gastric distress, numbness, and severe headaches, a common ailment for those times, and Mary attended him with great solicitousness. But a servant had seen her with the powder and warned Blandy that he was being poisoned. She even tasted the food herself and grew ill. Mary tried to destroy the powder by throwing it into a fire, but the servant rescued it. Nevertheless, quite soon thereafter, the elder Blandy died.

The local authorities were suspicious, so Cranstoun fled. Mary tried to follow him out of town but was arrested instead and sent to trial. No one knew quite how to prove a poisoning, but four doctors who had examined Blandy's internal organs at autopsy testified that the "preserved quality" of these remains was suggestive of arsenic poisoning. They believed that the white powder was arsenic as well, but could not prove it. One doctor applied a hot iron to the powder, which the servant had rescued, and analyzed it by smell. Neither of these "tests" was definitive, but both were state-of-the-art for the times. Mary admitted to using the powder but said she believed it was a potion to make her father more agreeable. Yet she could not explain why she had tried to destroy it.

The servant testified as well, describing what she had seen of Mary's behavior. The jury took just a few minutes to find Mary Blandy guilty of murdering her father. On April 6 of that same year, she was hanged.

In 1764, criminology—the psychological analysis of criminal behavior—also gained ground with the Italian publication of *Trattato dei délitti è elle pèna*, by Cesare Bonesana ("Beccaria"), reprinted in England as *Essays on Crimes and Punishment*. Beccaria found inspiration in the case of a Huguenot accused of killing his son for converting to Roman Catholicism. Under torture the man had confessed, but Voltaire, who had written about the case, believed that the boy's death had been a suicide. He included this analysis in a published plea for religious tolerance and his influence obtained a posthumous reversal of the conviction. He went on to try to stop the torture of religious dissidents. Beccaria set out to free the legal system from religious domination and regain some sense of classical ideals from ancient Greece, which based the state's power to punish in the collective welfare. Beccaria viewed crime as an offender's rational decision, so it made sense to devise a system of punishments that fit the crime.

John Fielding died in 1780, but by that time, Scotland had taken up the budding field of forensic science. A young woman was murdered on a farm, her throat cut open. The examining doctor looked closely at the wound and decided that a left-handed person had done the deed, since the slash started on her right. Investigators found a set of shoeprints that belonged to a man who wore footwear with iron nails, and the impressions indicated that he had been running and slipped in the mud. Near these impressions were drops of blood and a bloody handprint. Someone thought to preserve the prints with a casting material, which captured the impression of a recent mending job on the boot.

An autopsy showed that the dead girl was pregnant, but no one knew who her lover had been, not even her parents. Since it was a small village, the investigators believed that the man would have to attend the funeral, so they went to keep an eye out for a suspicious person. They also made each man there show the bottoms of his shoes and allow them to be measured. The authorities were thereby able to identify a man whose shoes had been recently mended and were the right size. He was also left-handed and had a scratch on his neck that had not yet healed. He said he'd been with two other men at the time of the attack, but the men recalled being near the cottage where the victim lived and that their companion had been absent for a short time. When he returned, they had noticed a fresh scratch on him.

When the suspect's home was searched, bloody stockings were found and an effort was made to match the mud on them to the mud from the bog where the footprints had been left. Indeed, the bog had quite distinct components, and the mud from the stockings matched. Then reports turned up that the suspect had been the dead girl's lover, and the case was complete. Finally, he confessed and was executed. It would not be long before Scotland established itself as a leader in forensic investigation.

Scholars in the field of toxicology continued to seek ways to improve the detection of poisons in human beings. The first breakthrough with arsenic poisoning was in 1775, when Carl Wilhelm Scheele discovered a way to change arsenic trioxide to arsine gas by treating it with nitric acid and combining it with zinc. Six years later in 1781, J. J. Plenck identified plant/vegetable poison in victims. In mid-decade, Samuel Hahnemann experimented with a three-step arsenic detection method that involved mixing it with nitric acid and heating it in combination with sulphurated hydrogen, yet the most significant discovery from that era was the "arsenic mirror." In 1787, Johann Daniel Metzger did some experiments and learned that when he heated arsenious oxide with charcoal, it formed a black mirrorlike deposit on a cold plate held over the coals. That substance proved to be arsenic. He managed to detect it in substances, opening the door for future scientists to figure out how to utilize this method in a forensic context. It seemed logical that if a death was a suspected poisoning, one could apply it to the stomach contents.

In England, Charles Dickens penned an article about the work of Scotland Yard, not yet an investigative agency, and he referred to the officers as "Detective Police." It appears to be the first use of the word *detective,* and Dickens utilized the traits and activities of several friends among the police as models for the detectives he created in his stories.

Another early incident that involved forensic analysis occurred in 1794, with a torn piece of newspaper. Edward Culshaw was shot in the head in Lancashire and the surgeon performing the autopsy removed a wad of paper from his bullet wound. It had been used to pack the shot, and the investigator flattened it out so he could examine it more closely. It appeared to be the torn corner of a piece of writing. When eighteen-year-old John Toms was identified as a suspect, his pockets were searched, which turned up a sheet of paper inscribed with a song ballad and missing a corner. This was laid against the bullet wadding, and the two torn pieces proved to be a match.

Toms was arrested, tried, convicted, and executed based on this early bit of objective analysis.

The following year in 1795, in the fourth edition of *Laws of Evidence*, a discussion was included on the subject of expert witnesses. The book noted the advancement of science, yet included the need for caution in light of physicians' mistakes. Just after publication, Franz Joseph Gall first devised the idea that moral character shows up in the shape of the skull, and he began to tour Europe to teach his ideas and acquire followers. He passed off as science a mistaken notion that would prove to be quite popular, to the detriment of people who merely had an unfortunate physical appearance. Yet even as Gall's ideas caught on, an impressive group of scientific pioneers during the next century would also trigger a spurt of growth in applying experience and experiments to criminal investigations. The Industrial Revolution had proven the value of the application of science to everyday life, and with this kind of encouragement, innovative minds quickly produced a number of marvelous inventions.

TWO

CHANGING SHIFTS

From Spies to Investigators

TOWARD SYSTEMATIC INVESTIGATIONS

As the 1700s came to an end, the University of Edinburgh had already thrown support to its practitioners of medical jurisprudence, and among them was Andrew Duncan, a professor of medicine who offered the earliest formal lectures on the subject in the English language. He even allowed members of the public to attend as a way to inject greater awareness into society. His aim was to urge physicians who participated in legal procedures to become more objective, systematic, and vocal in applying their knowledge of death and disease in the courtroom. The first formal chair in medical jurisprudence occurred in Edinburgh in 1807, with the establishment of the Forensic Science Institute. Duncan's son, also named Andrew, was its first occupant, and Americans seeking to improve their own haphazard field of medical jurisprudence looked to the Scottish physicians for guidance.

By this time, German-born Englishman William Herschel had discovered infrared light beyond the red portion of the visible light spectrum, inspiring Johan Wilhelm Ritter, a Silesian physicist, to look for invisible light at the violet end. With a prism and reactions to silver chloride, he thus discovered "chemical rays," or ultraviolet radiation, which would become a basis for identifying organic chemicals. Another Englishman, Thomas Bewick, viewed his fingerprints as identifying markers, but did not take the notion any further. Toxicology actually took the lead as a science applicable to criminal investigation when Dr. Valentine Rose showed in 1806 how arsenic could be detected in human organs. It was just in time for the sensational murder trial of a female serial poisoner.

Anne Schonleben neé Zwanziger, widowed at forty-nine, faced growing old alone and without money, so she hired herself out as a housekeeper in the hope of finding another man to marry. Her first target was Judge Glaser in Bavaria, who was married but separated. He seemed a good catch, but first Zwanziger worked to reconcile them to get access to the wife, whom she then poisoned. This left Glaser free to marry. However, he did not wish to marry Zwanziger, so she needed another plan.

She soon spotted a younger man who was merely engaged, and went to work on him. When he failed to respond to her, he died a ghastly death. Then another lawyer's wife died of gastric distress, which got Zwanziger booted from that house as well. By this time, she was getting desperate and feeling rejected, so she left some "gifts" behind. In the salt boxes, she placed some white arsenic powder, which made the servants, the master of the house, and his baby violently ill. However, they all survived and given the fact that they'd all grown ill at once, Zwanziger finally came under suspicion.

The police got involved, exhuming corpses believed to have been earlier victims. However, they could not prove poisoning beyond merely observing the state of the organs, which in the case of arsenic poisoning should be fairly well preserved. That turned out to be the case with these victims, so the authorities went in search of Zwanziger, already on her way to another potential "mate." They arrested her in 1809, finding a packet of arsenic on her person. She was held for trial, and just by luck, toxicology had taken a step forward just three years before when Valentine Rose, a member of the Berlin Medical Faculty, invented a method for detecting arsenic in the stomach or absorbed into the other organs.

Rose had cut up the stomach of a victim and dissolved it in water, filtering this substance many times before treating it with nitric acid, potassium carbonate, and lime to evaporate it into arsenic trioxide and treat it with coals. He thereby derived the telltale mirror substance that indicated the presence of arsenic.

Rose's procedure was applied to the organs of one of Zwanziger's victims, the wife of the judge whom she had initially targeted for a husband. It revealed arsenic, and along with the clearly identifiable arsenic in the salt containers of the surviving victims, the evidence against this predatory woman was convincing. Zwanziger then admitted that she had poisoned the judge's wife. She added the names of other victims as well, indicating that she had committed these murders for her own pleasure and had she not been caught, she would have continued to poison even more people. She called arsenic her "truest friend." In 1811, after being convicted of murder, she was beheaded.

Another linked set of brutal incidents drew attention in England, proving the need for a better police force. On December 7, 1811, the residents in London's East End near the Ratcliffe Highway docks had settled in for the night. One resident, Timothy Marr, was closing up his shop, assisted by his wife and apprentice. An intruder slipped in and murdered them all, as well as a baby in another room, by bashing in their skulls and slashing their throats. A constable who was called the following morning surmised from a bloody implement left behind that the weapon had been a seaman's maul with the initials *J.P.* set in copper nails into the handle. A small amount of money appeared to be missing, but otherwise the crime seemed to have no motive. The trick was to discover "J.P."'s identity.

There was no organized police force or investigative agents, so no one in the vicinity knew how to look for leads. The residents could only hope it might have been some ruffian passing through. But their peace of mind was quickly shattered again nearly two weeks later on December 19, when down the road a family residing at the King's Arms Inn was similarly slaughtered. The inn's owner, John Williamson, lay bludgeoned to death on the cellar stairs while the beaten bodies of his wife and the maid were in the parlor. The local residents demanded some form of police action, and circumstantial evidence soon pointed to an Irish sailor, John Williams. The constables searched his lodging house and found a trunk belonging to John Peterson (J.P.), a sailor out at sea, and the trunk was missing a maul. Based on a few witness reports, this slender "evidence," and the fact that Williams had no substantiated alibi or reason to be away from his room during the time of the second crime, he went to trial.

The courts of this era generally relied on circumstantial appearances, logic, and eyewitness testimony to form a narrative from the collected assortment of facts. When the pressure was on to convict someone so residents could feel safe again, as it was in this case, citizens might end up being scapegoated. But Williams was never tried, because he hanged himself first in the Coldbath Fields Prison.

For London officials, it was clear that with the growing population, they needed a way to focus on crimes and solve them. Leaving such matters in the hands of magistrates and constables was too haphazard. In addition, some of the local officials were open to bribes, so people with money might elude justice and continue with their crime sprees. The discussions began, but even so, an organized police force in London was still nearly two decades in the future.

Also in 1811, French pediatrician and chemist Pierre Nysten published the results of his studies of the state of a corpse in rigor mortis. He knew that at death, bodies initially went lax, but then stiffened up for a number of hours or days before once again returning to a flaccid state. He identified the various stages to formulate Nysten's Law, which stated that rigidity starts in the face muscles and descends from there to the lower limbs. He believed that progression had something to do with the distance from the brain. It was Nysten's hope that his "clock" would provide other physicians with an accurate estimate of the postmortem interval. While some of his reasoning was in error, his approach laid important groundwork for pathologists. In England, Dr. John Davey placed a mercury-based thermometer inside bodies to measure postmortem body temperature, adding algor mortis to the time-since-death indicators. However, neither man accounted for such things as environmental temperature or individual factors, so it seemed to them that they might be able to formulate calculations for an accurate postmortem interval and arrive at an exact time of death. Such precision eludes physicians to this day.

Now let's back up and return to developments in France, which would become significant for forensic science around the world.

UNDERCOVER

In 1804, Napoleon Bonaparte became emperor of France, shortly after America purchased the Louisiana territory from them, and he extended France's frontiers. By 1811, the French Empire ran from the Baltic Sea to the south of Rome, through Italy, Spain, Switzerland, and into most of Germany and Poland. Under Napoleon's encouragement, manufacturing and industry prospered, while poets, artists, composers, and novelists ushered in the Romantic Movement. Napoleon's exploits made good subjects for stirring tales. Dying for a cause became a noble goal and young men throughout Europe entered into revolutions in the name of glory and freedom. As an activist spirit spread, the Industrial Revolution, which put more money and goods into the hands of people, provided sufficient power to challenge authority. Yet while Napoleon was still in command in France, a former criminal, guilty largely of public scuffles and prison escapes, introduced an idea that would lead to a number of new inventions for criminal investigation, as well as inspire a fictional genre. To understand how he accomplished this, we must go back before Napoleon's time.

Born in Arras, France, on July 23, 1775, François Eugène Vidocq was the son of a baker, but from a young age he viewed himself as an adventurer. Toward that end, he ran away several times, including an attempt to reach America. However, before he could board a ship, a clever prostitute robbed him, leaving him without means to go anywhere, so he sullenly returned home. Taking out his frustration in swordplay, he killed a man in a duel. As punishment, he received a choice: go to prison or join the military. King Louis XVI and Queen Marie Antoinette needed fighting men to help resist the movement for a new republic, so Vidocq agreed to serve in the Bourbon Regiment. But he could not keep himself out of trouble.

On leave in 1794, after Louis and his queen had both lost their heads on the guillotine, Vidocq interrupted an execution that disgusted him, and was arrested. A high-placed official intervened on his behalf, but Vidocq became too familiar with the man's daughter and found himself married and in charge of a grocery store. This was hardly to the liking of a self-styled adventurer in the last breath of his adolescence, so when the girl proved faithless, he returned to his more aggressive career path. In Brussels, Vidocq posed as a military officer and became a successful cardsharp before being arrested yet again, this time for supposed desertion. He escaped and returned to Paris, where he made a violent scene over his mistress and another man, and was promptly arrested there, receiving a three-month sentence.

During his first actual prison stint of a serious nature, Vidocq performed an act of kindness that was to change his life, for both the worse and the better. He met a farmer serving six years for stealing grain to feed his family. Taking pity on the man, Vidocq forged a pardon for him. When prison officials discovered the deception, they slapped Vidocq with a harsh sentence of eight additional years. Refusing to capitulate, he escaped, was caught, and escaped again. He seemed always to find ways to charm or elude officials, but eventually he would make a mistake and end up back in prison. At one point, his reputation for breaking out was so great among the jailers that they forbid him from even leaving his cell. To their surprise, he managed to rig a disguise that got him past them. Yet for all his escapes, he kept getting caught and his sentences increased in severity until he ultimately received a life sentence in a prison so brutal that many men who went there buckled and died.

While Vidocq was busy moving from one arrest to another, Napoleon Bonaparte took charge of the army and declared himself the master of France. He instituted a new code of laws, expanded the empire, and granted the citizens greater personal freedom. Invading other countries with flourish, Napoleon made Paris the capital of Europe and rebuilt it into a magnificent new city of monuments and boulevards. Yet these improvements also attracted criminals, and since the police force focused largely on political subversives, the crime rate for theft, forgery, and murder climbed, which would one day prove beneficial to Vidocq.

In another part of France, he was still attempting to regain his freedom, but this time by appeal to reason. While romantic stories abound as to how he eventually managed to join law enforcement, the truth appears to be more mundane. Still, one tale has become so firmly a part of his legend that it would be remiss not to tell it. Supposedly, Vidocq asked the prefecture of police (the specific official and city change with different authors) for the chance to prove he wished to go straight by becoming an informer on the criminal element. He knew these people, he pointed out, so who better to go among them to learn their secrets? The prefecture said he had no choice but to send Vidocq to prison, since he had not served out his sentence, so Vidocq posed a deal: If he should succeed in escaping and returning to the prefecture's office, that very act should sufficiently prove his sincerity. That scenario seemed too improbable for the prefecture so he agreed, and then added extra guards to escort Vidocq to prison. Yet Vidocq was soon back in the official's office, ready to begin his new occupation.

In a less imaginative account, Vidocq explained his plight to Jean-Pierre Dubois, chief of police in Lyon. Dubois looked into it and realized that Vidocq was actually guilty of only a minor misdeed, so he offered the deal to become a police informer. Grateful to be free of prison, Vidocq accepted.

No matter which tale of entrée is true, Vidocq eventually went among criminals who knew him as one of them, learned what he could about planned criminal activities, and duly reported them to Dubois. Vidocq succeeded quite well and the arrest rate improved, yet the criminals could not figure out how the police knew so much. Eventually Vidocq moved to Paris with his mother and mistress, and in 1809 continued his activities as an informant for the criminal division of the prefecture of police. How that came about presents yet another mythical narrative that added to Vidocq's mystique.

It seems that the empress Josephine discovered that a priceless emerald necklace, a gift from her husband, was missing. Napoleon ordered his director of police to spare no effort to find the necklace. However, the police had no training in looking for thieves and no network of informants, so they did not know what to do. Vidocq stepped in, donning a disguise to enter the taverns where rogues gathered so he could acquire information about such a theft. He listened to others brag about their plunder and soon had the leads he needed. Within three days, so the legend goes, Vidocq managed to locate the necklace and deliver it safely, as well as bring the thief to justice. It was reportedly this feat that got Napoleon's attention, who then urged Vidocq to continue his work as an undercover informant.

In whatever manner he actually achieved his new occupation, underground he went. Officials "sent" him to La Force prison, where he mingled among the prisoners, making reports twice a week about what new criminal activities to expect on the streets. He even solved a few unsolved murders by listening to his "fellow" convicts discuss their deeds and make arrangements with one another for more crimes. One of his first tasks was to get evidence against "Coco," a notorious burglar who had stolen from a police official and who would shortly be up for trial. Vidocq ingratiated himself with the man, learned that the police had failed to question the only witness, and delivered the name of that person. To Coco's surprise, he was convicted.

After a spell in the prison, Vidocq requested a street assignment and was taken out in irons so he could "escape." He merged into the Parisian underworld, a hero among them for his ability to "defy" the authorities. Since these men believed that Vidocq had actually escaped, he won their respect and their confidences, which he quickly betrayed. He was responsible for hundreds of arrests.

Vidocq's growing success eventually won him an audience with officials, to whom he proposed a whole agency staffed with undercover informants like himself. They would be plainclothes agents who could move freely across the city's many jurisdictions rather than being "turfed" like the gendarmes were in specific districts. It seemed a good plan, at least as an experiment, so he received permission to hire four men. They worked out of a three-story building in which Vidocq could store his innovative card files and police reports, which grew exponentially, and the force eventually expanded to eight, then twelve, including a few women. In October 1812, Napoleon signed a decree that turned Vidocq's *Brigade de la Sûreté* into a national security force. Thus, Vidocq created the world's first undercover detective organization.

He taught his cohorts a number of clever moves, particularly with disguises. As one story goes, when a gang of Parisian criminals wondered who among them was the informant that had lately triggered an increase in arrests, no one suspected the sailor with the month's growth of beard, the redheaded pirate, or even the scar-faced, dark-haired gypsy. Yet they were wrong on all counts. Each of these supposed thieves or grifters was in fact a single person: François Eugène Vidocq, a master of disguise and the very informant they sought to flush out. Supposedly he even accompanied a group of assassins who planned to ambush and kill Vidocq, but failed when their target did not show up.

To blend in effectively, he not only changed the color of his hair and the outfits he wore—including becoming a woman—but also advocated that to be successful at duping others, one must be fully immersed in the character one adopts, including special effects such as dirt, sweat, and shabby clothes. He supposedly once summed up the secret to passing as someone else for the author Honoré de Balzac: "Observe what you would become, then act accordingly and you will be transformed." Vidocq's own favorite personas were "Jean-Louis," a sixty-year-old "fence" who paid high prices and spoke like a Breton, and "Jules," a bearded burglar who preferred physical force and was frequently seen with his blond mistress (another informant).

In the meantime, politics in Paris took a new turn. French domination of Europe had produced heavy taxation and trade restrictions that fed a growing desire abroad for Napoleon's demise. Eventually Britain, his enemy since the War of 1812, which set the boundary between the United States and Canada, invaded France. Napoleon also took a blow from Russia, and his humiliating retreat inspired other countries to move against him. By 1814, Paris had fallen and Napoleon was banished. He attempted to regain his throne, but the British and Prussian armies defeated him at Waterloo. Yet France was treated well at the Congress of Vienna in 1814–15, and its royal rule was restored.

The *Sûreté* continued with business as usual, and its success brought respect to law enforcement in Paris. Branches were soon established in nearby towns. As per Vidocq's philosophy that a good job can transform a criminal, the agents all had criminal records and only Vidocq knew their identities, which annoyed the uniformed police. Despite doubts from officials about trusting former cons, these men and women quickly proved themselves, making Vidocq proud.

Teaching his associates that "observation is the first rule of investigation," Vidocq devised a system of clever techniques for acquiring information, especially with interrogation. He also offered a form of plea-bargaining, and he gained a reputation for being fair. Yet he was not above using tricks. When he went to search a place, he might tell a suspect that he was being investigated for a different crime than the one for which he was suspected. That person, knowing he was innocent of *that* offense, would allow Vidocq to search his rooms, and Vidocq would then look for evidence of the actual crime. (No one worried then about a criminal's rights.)

Vidocq's focus on a systematic approach to crime helped him to develop some of the forensic techniques in use today, such as keeping detailed written records, comparing spent bullets to weapons (mostly for size), preserving footprint impressions with plaster of Paris, comparing samples of handwriting to forged notes, and suggesting that fingerprints might be used as a form of identification. Vidocq also worked with area chemists to create forgeproof paper and indelible ink, and he took out patents on both. Whenever he had a body removed from a crime scene, he required the "most minute exactitude" to items at the scene, not knowing when some slight scrap of paper or a button might become an important clue. If a robbery occurred, he would look for people in the underground who suddenly spent money in an uncharacteristically free manner. Thus, he might just as easily be called one of the earliest criminal psychologists.

In 1833, Vidocq submitted his resignation from the *Sûreté*, but he did not retire from law enforcement; he had too much going for him to call it quits. Early the following year, he established *Le Bureau des Renseignements*, the world's first private detective agency. Given his reputation across Europe by that time, Vidocq had no trouble attracting clients who wished to locate confidence men who had stolen their money. His agency took on more personnel, utilizing the methods and tools he had already perfected, and he remained active until he was eighty years old. He died two weeks after suffering a stroke, just short of his eighty-second birthday.

Despite the resistance among established law enforcement to Vidocq's "new" methods, he made such an impression and was so successful that he inspired many crime novelists—and thus, became the archetype of the forensic sleuth. Throughout his life, his exploits were legendary and when he submitted his memoir to a publisher, a ghostwriter beefed it into a bestselling sensation in 1828—albeit largely exaggerated and untrue. Vidocq was unhappy with it, but he did not seem to mind how his literary friends used his life and exploits for their own gain. He was the role model for Balzac's clever Vautrin, and he inspired characters in novels by Victor Hugo and Alexandre Dumas, père. All three authors were Vidocq's frequent dinner companions, as were the prominent literary critics, social reformers, and poets of his day. Several stage plays in England, where Vidocq was highly regarded, were based on his memoir. He even penned a few novels himself (or allowed his name to be used), thereby giving the world yet another innovation, the first detective story (beating Edgar Allan Poe by a few years).

During Vidocq's years as a detective, a steamship crossed the Atlantic on its own power, which presaged an independence from the forces of nature. Industrialization swept through Europe, which gave greater support to the areas of science that could boost production and make daily life more efficient. The United States laid tracks for a network of cross-country trains, while waves of emigration to new lands took European culture and technology onto other continents.

At the same time, ideas formed that benefited criminal investigation, and dentistry received attention. Frenchman Pierre Fauchard had grounded the discipline with his comprehensive 1728 publication, *Treatise on the Teeth,* and dentists around Europe who read it shared a common knowledge, aware that teeth remained constant and had characteristics that could help to identify the dead. Thus, some dentists found themselves invited into criminal cases. In 1814 in Scotland, a medical lecturer and two of his students from Glasgow were charged with violating the grave of a Mrs. McAllister during an attempt to acquire bodies for dissection (a common but illegal practice). Prosecutors had to establish her identity, so Dr. Alexander, her regular dentist, testified that he had constructed a partial upper set of teeth for her. He took a mold of the teeth he had made and demonstrated how he could fit the denture in a skull, which the relatives of the deceased had identified as Mrs. McAllister's. Another dentist affirmed the fit, and all seemed well until the defense attorney produced his own expert. This dentist stated that the upper denture did not articulate with the natural teeth in the mandible and argued additionally that artificial teeth could fit other jaws equally well. Since the professional testimony conflicted, producing a reasonable degree of doubt, the court made a finding of not guilty. Once again, people questioned the concept of a scientific expert.

In 1823, Czech physiologist Johannes Evangelist Purkinje published a description of nine fingerprint types, thinking along the same lines as Vidocq had that fingerprints might prove to be the basis for identification, and Thomas Bewick, an English naturalist, used his own fingerprints on his published books. In 1828, William Nichol invented the polarizing microscope, the French worked on methods for bloodstain detection, and John Glaister was putting together his studies for *Hairs of Mammalia from the Medico-Legal Aspect.* Toxicology, however, was about to make the most dramatic leaps forward.

CHEMISTRY, BULLETS, AND BLOOD

At the Paris School of Medicine, Mathieu Joseph Bonaventure Orfila received his medical degree. Originally from Spain and a former child prodigy, in only two years he became the most eminent name in toxicology. In

1813 at the age of twenty-six, he published the first systematic treatise on known poisons, *Traité des Poisons . . . ou Toxicologie Générale.* In this, he documented and organized the findings of scientists before him to offer a systematic guide to everything then known. He had set up a rudimentary lab in his own home, doing experiments there on animals and offering private lectures. It was when he found some methods failing to produce what he expected that he set about to learn why. That's how he moved the chaotic field of toxicology into the true mode of science.

Four years later, Orfila published a second book, which earned him an appointment as professor of medicinal chemistry at the University of Paris and renown throughout Europe as the foremost authority on poison. It was Orfila's experiments that showed how arsenic traveled from the stomach to other areas of the body, so physicians now knew that its absence in the stomach was no proof that a person had not been poisoned.

Despite his fame, it was not until 1824 that police consulted with Orfila on a criminal case. A man had died suddenly, within ten days of getting married, and an autopsy indicated that there might be arsenic in his body, so his new bride was arrested. The contents of his stomach went to Orfila for tests, but he found no arsenic, and since that was all he'd received to work with, the widow was acquitted. As his fame grew, Orfila repeatedly found himself in a battle of experts in court as toxicologists on the other side went up against him. His careful approach won him greater respect, so his opinion often prevailed.

Nevertheless, this type of clash was an ominous sign of what lay ahead for medical jurisprudence. If toxicology and medicine claimed to be scientific, i.e., precise, then the results of these so-called objective tests ought to be the same for anyone. That experts could end up on opposite sides, with both experts confident of their findings, undermined for the public the claim that they offered indisputable facts. It seemed more subjective than they admitted, and even open to being purchased by the attorneys. Medical jurisprudence throughout Europe and America would suffer over this issue into the next century.

Yet toxicology did find success in an American case in 1828. In a village near Philadelphia, a man died following an illness, and Dr. Samuel Jackson, a medical educator and expert on pharmacology, considered it the effect of heavy drinking. However, people who knew the man believed that his wife had poisoned him after she'd allegedly purchased arsenic. The local coroner had four doctors review the case, and they conducted primitive tests for arsenic. They concluded that the man had been poisoned, so Jackson came back into the case to point out the French method of having everything double-checked by a reputable expert, and he upbraided the doctors for not sending the viscera for testing in Philadelphia. When the grand jury voted not to indict, Jackson was pleased that the reasoning of science had trumped a loose collection of innuendo and gossip.

American physicians noticed that resolving a case that garnered publicity was a road to fame and professional advancement, and people did attend trials for entertainment. It was a productive venue to demonstrate what medical science could do. The doctors in America looked to noted figures such as Orfila and aspired to gain that sort of reputation for themselves. One such person was Samuel Gross, who had taken an early course in medical jurisprudence in Philadelphia. He lived in Easton, Pennsylvania, and as luck would have it, ended up in a sensational case in 1833. A woman eight months pregnant was strangled to death, and the suspected culprit was her lover, a man named Goetter. Gross performed the autopsy and thus came before the court. But Goetter had engaged the service of a famous Philadelphia lawyer, James Madison Porter. When Gross took the stand, Porter attacked his vulnerability, notably that he had failed to examine the woman's brain for signs of a cause of death other than strangulation.

However, Gross compensated for his error by describing the research on signs of asphyxiation by strangulation. Indeed, he had done his own research by strangling small animals and dissecting them to examine the damage done to the tissues. Despite the defense's display of a dozen physicians brought in to contradict young Gross, he stood his ground. Since the circumstances went against Goetter, he was found guilty. When he confessed just before he was hanged, Dr. Gross achieved the fame he sought, along with professional advancement, and the incident affirmed the scientific method.

Investigation was making some progress in England, too. In 1829, the British Crown assented to Sir Robert

Peel's Metropolitan Police Bill, which was the beginning of an organized, full-time police force in London. Called Peelers or Bobbies (both derived from Peel's name), they were housed in Whitehall Place near Great Scotland Yard (and later moved to another building called New Scotland Yard). During the 1840s, a separate detective department was established, and over the course of the next few decades, most of Britain worked on replacing parish constables with local police forces.

Early in 1835, Bow Street Runner Henry Goddard investigated a case of an alleged burglar who had invaded a stately home. The home's butler, Joseph Randall, reported that a masked man carrying a lantern had awakened him and he'd described the lantern casting a shadow in front of the intruder, who was accompanied by a second man. They quickly left but fired into the room on their way out. The bullet shot past the butler's head and lodged in the bed's backboard. Randall ran out, encountered the men again, and scared them off, rescuing the goods they had attempted to steal.

Goddard listened to this story but found it less than convincing: The way the would-be burglars had entered made little sense, especially when he noted marks on the inside of the door frame they had supposedly pried open. He also examined a closet in the home they had allegedly pried, and the marks indicated a different tool. Why, Goddard wondered, would they change their method in the middle of a job?

Hypothesizing from inconsistencies between Randall's story and the evidence that the burglary may have been an inside job, Goddard asked to see the butler's gun and the bullet fired at him in his bedroom. When these were produced, Goddard thought that while deformed from impact, the fired bullet had characteristics that were similar to the bullets the butler used in his gun. Each had a tiny imperfection, most likely produced by the mold used to make them. Goddard asked the butler for the mold and quickly located the source of the imperfection—a tiny hole. He took all of this to a gunsmith, who confirmed that the mold had produced the fired bullet.

Confronted with these findings and with the different types of pry marks, the butler confessed that he had staged a burglary and shot into his bed board himself. He'd been hoping for a nice reward from the grateful mistress of the house for saving the goods from the "intruders." With the resolution of this case, Goddard became the forerunner of a long line of forensic ballistics experts, although it would be decades before the examination of projectiles would become a rigorous discipline.

English chemist James Marsh made the next dramatic advance for toxicology. Although at the time arsenic could be detected in the organs of victims of poisoning, there was no known way to measure its quantity, and defense attorneys were suggesting other ways that arsenic could enter the body. Wallpaper paste, for example, or hair products and even women's makeup often contained arsenic as well. In October 1836, Marsh published an article in the *Edinburgh Philosophical Journal* to describe how one might measure small quantities in such a way as to indicate ingestion and absorption. He had developed his method from a rather peculiar case, which had forced him into months of persistent analysis.

George Bodle had died after drinking coffee and his symptoms, along with the reportedly strained relationship he had with his family, suggested that he may have been poisoned. According to the reports offered, on the morning of the death, Bodle's grandson, John, had filled a kettle with water from the well, a behavior that the maid reported as uncharacteristic. He'd also previously expressed to acquaintances a desire for the old man to die. The kettle had then been used to brew the coffee.

Marsh tested both the kettle and the coffee that was made that morning, and he found the presence of arsenic in both. However, when he testified in the case before a jury, he had a difficult time describing the test he had used. His scientific explanation proved too abstract for the jurors so they declined to convict. (A decade later, the young suspect would confess.)

Frustrated over his inability to communicate with a jury, Marsh set about making his methods more demonstrable to uneducated minds. Should he be invited into another case, he decided, he did not wish to watch all his work go to waste. He knew that Metzger's test could show how arsenic formed the mirror deposit, but since the arsine gas escaped into the air, allowing small amounts to remain invisible, Marsh sought a way to contain it. In a sealed bottle, he treated poisoned material with sulphuric acid and zinc. From this bottle emerged a narrow U-shaped glass tube, with one end tapered, through which the end product would emerge into a flask containing zinc and sulphuric acid. If there was arsine gas present, it passed into a heated glass tube and could be ignited to form the expected black mirror substance in a cooler area of the tube. Marsh did manage to use the test again, and it proved sufficiently visual to illustrate his explanation. The method, which could detect even miniscule amounts of arsenic, became known as the Marsh Test.

In France, Mathieu Orfila recognized the Marsh Test's value and he used the device to test for arsenic in exhumed bodies and in cemetery soil, so that he could show whether a suspected poisoning victim had more arsenic in his system than was consistent with the amount in the surrounding dirt. However, he concluded erroneously that arsenic could not enter bodies from the soil if they were buried in coffins.

CELEBRITY CRIMES

As a few experts and investigators gained fame, the public grew more fascinated with criminal cases. Sometimes they rooted for the noted authorities but just as easily supported a defendant. While newspapers covered these cases and brought greater awareness to the feats (and foibles) of forensic developments, certain traditions that evolved into tabloid journalism arose from serendipitous decisions made during a case in Manhattan.

Mrs. Rosina Townsend's bordello of nine girls experienced a fire early in the morning of Sunday, April 10,

1836, and one girl, Helen Jewett, was found murdered in her bed. The key suspect was Richard P. Robinson, nineteen, a cultured young man and one of Helen's frequent clients. He claimed innocence but certain pieces of evidence implicated him so he was bound over for trial.

During the investigation, James Gordon Bennett, editor of the *New York Herald*, went to the crime scene to view Jewett's body and interview Townsend. He observed how the subjects of sex, crime, and scandal would be edgy, and he was willing to endure criticism to make his reporting more interesting. He succeeded at both, and the popularity of his coverage was so great that people in other cities demanded to know more about this murder, so for the first time journalists traveled to report back to their own papers.

In front-page articles, Bennett described the sensuality of the victim's body under the sheet that was placed over her. He suggested that Robinson might indeed be innocent, while other papers sided together against the *Herald*'s coverage and offered sympathy to Helen. A cult movement developed in which young men supported Robinson with cloaks and caps like his, and attitudes that men had the right to do what they wanted. Yet Helen, too, had supporters—women who donned white beaver caps.

Then Robinson's journal was found in his room in which he described how his innocent looks belied the depravity of his soul. The papers now identified him as a "consummate scoundrel," while Helen became a beautiful girl seduced.

More interesting was the physical evidence against Robinson. He admitted that he owned the cloak left behind in the bordello that night, and a broken piece of twine attached to a buttonhole of his clothing appeared to coordinate precisely with broken twine on the hatchet handle used to bludgeon the victim. A porter from the store where Robinson had worked identified the hatchet as the one Robinson always used. It had turned up missing on the Monday after the murder. The twine on Robinson's coat also proved consistent with the type of twine used at the store, and Robinson's roommate could not alibi him. There were traces on his trousers of whitewash, apparently from the backyard fence over which he presumably had climbed leaving the bordello. He'd also shown no surprise when the police came to question him. Letters back and forth between Helen and Robinson indicated that they had been quite attached and he was jealous. When Helen learned of his plan to marry a respectable girl, she threatened to publicly humiliate him. Robinson sent a note saying he would come to her but asked her not to tell anyone about his visit. He would be there on Saturday night—the night of the murder. About a week before the murder, a man named Douglas had attempted to purchase arsenic at an apothecary close to the bordello for getting rid of rats. The clerk had refused his request, and in the courtroom that day, this same clerk provided a dramatic moment by identifying Robinson as Douglas.

But at the June 1836 trial, Robinson's defense attorney had his own dramatic moment as he presented a handkerchief taken from under Helen's pillow to show that Robinson had not been the only person in her room that night. A witness, Robert Furlong, provided an alibi—that Robinson had been in his store that evening smoking "segars" and reading. The supposed whitewash stain on Robinson's trousers was consistent with paint from the store where he worked, and a manufacturer of the hatchet testified that he had sold some 2,500 in New York City. Thus, each item of the circumstantial evidence was reinterpreted in Robinson's favor (and probably with the help of a bribed witness).

For the first time in American history, representatives from newspapers in other cities were present. On June 8, the jury found Robinson not guilty, which caused one side to be angry while the other felt vindicated. After this case, thanks to keen public interest, journalism showed greater interest in covering crimes, especially those with a salacious side.

Two years later a man was murdered in Baltimore, his face cut so much with a hatchet it was difficult to identify him. He'd also been shot, and the bullets removed from him proved to have been specially made for William Stewart, who had recently claimed his father's inheritance. The victim, it turned out, was Stewart's father, and thanks to the identification of the bullets, Stewart was convicted of murder.

Around this time in Boston, the first professional American police force was established as the "day watch." Based on the British system, it diverged by retaining local instead of central governance.

MORE POISONERS

In 1839, the word *scientist* came into use, H. Bayard published the first reliable procedures for the microscopic detection of sperm, and photography was invented in France. Louis Jacques Daguerre and Henry Fox Talbot developed the process and a young writer named Edgar Allan Poe, who would soon write his first detective story, penned several essays about the daguerreotype. It involved a mirror-polished silver-plated copper sheet, treated with iodine fumes, which converted its surface into a thin coating of silver iodide. After the plate was exposed in a camera, it was developed with a vapor of metallic mercury. Poe noted its superiority to mere description of a person, and even as cameras were utilized for portraits, police officers saw their value for mug shots.

As photography evolved, toxicology suffered a setback in a notorious British case. People reading the papers believed that Wainwright the Forger, an artist and former contributor to *London Magazine*, had killed three relatives and a friend for profit. Yet because he had allegedly used strychnine, a poison difficult for toxicologists to detect, these scientists were unable to prove the murders. Wainwright was sentenced for forgery and sent to Australia, but the populace decried the scientific findings and believed that he had gotten away with murder.

That same year in Paris, Mathieu Orfila entered a notorious case. A gendarme named Dupont wanted to marry the daughter of a man named Cumon, who lived in Montignac. Cumon refused the request, since Dupont had few prospects for taking care of Victorine. But like Mary Blandy, Victorine most ardently desired to make the match so she continued to encourage Dupont, which precipitated quarrels with her father. Victorine cried on the shoulder of her nursemaid, Nini, and before long, the old man fell ill and died from a gastric condition. Victorine, who seemed to bear this loss without much sorrow, immediately married Dupont. Given the circumstances, Cumon's death was clearly suspicious, but no one investigated.

Then Nini heard an emotional speech made by a condemned convict who had passed through the village, and in remorse she admitted that she had purchased the arsenic and other toxins that had rid her mistress of the hindrance to marrying Dupont. She added that Victorine herself had administered the fatal substance. Cumon's body was exhumed and submitted to experts. The local doctor said there was no evidence of wrongdoing, but Orfila had relied on the Marsh Test. His demonstration was apparently more convincing than the doctor's word, so Victorine was convicted and received life in prison, while Nini got eighteen years of hard labor. Largely due to cases like this, Orfila's name became established, and by the following year, he had to rescue the field of toxicology from bunglers.

Twenty-four-year-old Marie Lafarge had been arrested at Le Glandier, France, for the murder of her husband,

Charles. Known to be unhappy with her arranged marriage to the owner of a rat-infested forge, she stood accused of using arsenic planted in food as a murder weapon. She had bought a relatively large amount during the months preceding the death, allegedly to exterminate the rats, and her husband had become violently ill from eating a piece of cake. Servants claimed that Marie had fed him herself and had stirred white powder, which she claimed was a special type of sugar, into a glass of milk. But a servant noticed white flakes remaining on the glass and showed them to a local pharmacist. He tested the food and found arsenic. Lafarge's doctor, too, had suspected someone had been poisoning Lafarge over the past month. The circumstances were clearly against her.

However, because Marie had aristocratic connections and proclaimed her innocence, the populace grew interested in her case and sided with her. Marie's tale of woe involved being married to a coarse and foul man who had lied about his wealth and had married her to get access to her family's money and her dowry. She had never wanted to marry him and had begged to be released. She proved to be a romantic figure and was soon a national celebrity. A handsome young attorney became her advocate and she presented herself as a graceful and poised defendant.

The prosecutor portrayed her as a cunning monster, but his experts were unable to determine with the Marsh Test that the contents of Lafarge's stomach contained arsenic, so they requested that the body be exhumed to test the organ tissues. Still, the tests were negative. Given the circumstances and the testimony from Lafarge's servants, this result seemed impossible. Food from the Lafarge household was tested and revealed the presence of arsenic. The experts were stumped.

Marie's attorney, M. Paillet, invited Orfila to review the experts' work and conduct his own tests, obviously hoping to prove the prosecution wrong. But Orfila noted the obvious carelessness with the procedures. He examined the materials the experts had used for their analyses and noted their clumsy reliance on the primitive Rose test. This proved their lack of awareness of the progress made in toxicology. That was a score for the defense. But the judge allowed the prosecutor to have his experts use the Marsh Test. Again, the country waited in anticipation of the results, and again, the results turned up negative in the body. People who sided with Marie rejoiced over her triumph, but there was still the issue of large amounts of arsenic evident in the food.

The prosecutor then requested that Paillet's friend, the renowned Orfila, perform the tests on the contents of Lafarge's stomach. Orfila arrived and insisted that the local chemists witness what he did. He used the Marsh Test, and to Paillet's shock, he proved that it was not the method that had been at fault in this case, but the practitioners. They had bungled it. Orfila tested many of the organs from Lafarge's body and detected the presence of small quantities of arsenic. He also proved that it had not originated in the soil surrounding the coffin, since arsenic was not present in the flesh. Based on his results, Marie was declared guilty and sentenced to death. Her sentence was later commuted to life and she spent ten years in prison. She never confessed.

Since Orfila had been Paillet's friend, this case offered a demonstration that benefitted toxicology. Orfila had performed the test in a detached manner, uninterested in the outcome aside from accuracy. He might have chosen sides, but he did not. He also showed the reliability of the test and the importance of careful, systematic work.

More devices would soon acquire application in criminal investigation, as the century moved toward its midpoint and the Industrial Revolution continued to transform life in every major European and American city.

THREE

ORDER OF THE COURTS

Judges and Juries

NEW TECHNOLOGIES

In 1840, Samuel Morse obtained a patent for the telegraph, and four years later he sent a message from Baltimore to Washington, D.C. Other countries adopted this exciting means of instant communication and it wasn't long before an underwater transatlantic cable linked Europe with America. In England, an investigator quickly employed this invention to apprehend a murder suspect.

John Tawell was seen leaving his home on New Year's Day, 1845, and arriving at the home of his mistress, Sarah Hadler, in Saltill, England. As evening set in, a neighbor who'd seen Hadler looking quite pleased during the day heard her let out a piercing scream. This neighbor then saw Tawell run off so he went to check on Sarah, who lay on the floor of her cottage, writhing in agony. She died before the doctor arrived.

Tawell, married twice, had already served a prison term for forgery, and had taken up with Hadler during his first wife's illness. When she died, he'd begun seeing Hadler more regularly. Yet he'd married another woman and kept Hadler, with whom he had two children, a secret. He was the obvious and only suspect in Hadler's apparent murder, so to catch him before he got away, the local constable in Hadler's village telegraphed a message to the town where Tawell was expected to arrive, and there he was spotted and caught. Hadler's death was proven as a poisoning by prussic acid, which Tawell had purchased. However, in court, the toxicologists made a mess of things by contradicting each other on how much prussic acid apple pips contained, since apples had been present in Hadler's home. The jury ignored the medical evidence and convicted Tawell based on circumstances. He later confessed.

In 1843, the Belgium *Sûreté Publique* took the first known mug shots of criminals, making this group of law enforcement officers the ancestors of judicial photography. The Swiss used the technique during the next decade for identifying suspected recidivists. Throughout the 1850s, police departments across Europe and the United States also compiled archives of prisoner images.

The early psychiatrists, who were communicating with others in their field via medical journals and papers, sought ways to identify criminals and treat them apart from the nonviolent mentally ill. By mid-century, while they had no tool for systematic diagnosis, these psychiatrists had established the notion that a person who seems unaware of doing something wrong when he or she commits a crime should be viewed as medically insane. Because some of them seemed obvious by appearance, the fashion, coupled with physical anthropology, was to develop a way to use the obvious for making a diagnosis, but to offer this as a controlled result of observation and measurement. The earliest attempts looked to the shape of the head.

CRIMINAL INSANITY

With the rise of modern science and the emphasis on natural law and material substance, the appraisal of human character from external appearances became a fashion. Phrenology, first proposed by Austrian physician Franz Joseph Gall in 1796, espoused that the brain was the organ of the mind and thus the basis of personality. Gall stated that moral and intellectual faculties are innate, they're locatable in specific areas of the brain—"organs"—and the form of the head represents the form of the brain, which reveals the development of the various areas. Gall called his idea organology, but later in Britain, physician T. I. M. Forster coined the term *phrenology*. Gall's colleague and disciple, Johann Spurzheim, also referred thus to the practice.

Phrenology involved feeling the various bumps or depressions on a person's skull to determine how different areas of the brain were functioning. It was dividable into thirty-five distinct organs, each associated with a specific trait such as "secretiveness," "firmness," "adhesiveness," and even "philoprogentiveness." The larger the area of the organ, as evidenced by a bump in the skull, the more pronounced the trait was believed to be.

When phrenology filtered into mainstream society, the phrenologists became matchmakers and employment advisors as people consulted them on everything from the best qualities for a romance to whether they should hire a certain person. Yet it was not accepted as a science in quite a few circles, notably in Edinburgh, largely because there were many counterexamples for which the phrenologists failed to account, so its proponents had to fight to have their work taken seriously. Eventually they gained a foothold, even in Edinburgh, where the first phrenological society was founded. Although the British Association for the Advancement of Science excluded phrenologists, their own journals and associations were modeled on the practice. This approach to understanding and predicting behavior proved highly popular in America, especially toward the middle of the century. It also spawned other approaches with a similar notion that physicality was key to personality types. Hubert Lauvergne, a physician at Toulon Penitentiary, observed that many convicts had unusual faces, which he believed must reflect their criminal instincts. For a time, prisoners would be classified according to their phrenological profiles—a listing of traits specific to them, based on skull formation.

Brain damage supposedly caused insanity, and theorists set about distinguishing the mentally deficient from the deranged. Autopsies were often performed on mental patients to locate affected areas of the brain that would account for some trait or behavior. The superintendents of mental institutions during the nineteenth century, called alienists from their expertise with *aliéné* or mental alienation, founded the rudimentary practice of psychiatry and encouraged one another to report interesting cases.

Despite the growing emphasis on free will operating in criminal behavior, there were clear instances in which someone acted from a delusion or uncontrollable impulse, and some experts who studied them were pleading for greater understanding. In 1837, Matthew Allen published an essay on the classification of insanity. He discussed the role of stressful relationships as a precipitator and described one of the first cases of parole into an asylum. In that same year, American psychiatrist Isaac Ray published a tract, *Medical Jurisprudence of Insanity*, which laid the basis for forensic psychiatry. Shortly thereafter, in 1843, the issue of insanity challenged a British court.

Sir Robert Peel, founder of the English Bobbies, was the target of a deranged man who mistook his secretary, William Drummond, for him. Daniel M'Naghten believed that "Tories" were conspiring to murder him, so he shot Drummond as a preemptive act of self-defense. Since M'Naghten had a history of such delusions, his attorney found plenty of support for a defense of madness. His counsel claimed that he had not known what he was doing and was not in control of his acts. The prosecution offered no rebuttal, acknowledging the obvious, so the court declared that the accused must be found not guilty by reason of insanity. M'Naghten was detained in an asylum to keep him from harming anyone else.

However, since he'd shot a public official, British citizens reacted with anger to this finding, so a royal commission undertook to study the issue. They examined every angle and agreed with the court, formulating a response that became the M'Naghten Rule, and thereafter the House of Lords required that to establish a defense on the ground of insanity, it must be proved that "at the time of the committing of the act, the party accused was laboring under such a defect of reason, from disease of the mind, as not to know the nature and quality of the act he was doing; or if he did know it, that he did not know he was doing what was wrong." The legal system in the United States likewise based such decisions on this ruling.

Throughout the inquiry, there was no mention of M'Naghten's phrenological profile, but phrenology still had firm adherents. In 1845, German Gustav Zimmermann wrote that if phrenology could accurately identify the marks of "villainy" on a person's skull, every person age twenty-five and over should submit to a phrenological examination. If they were found to have a dangerous predisposition, they should be hanged or confined so as to prevent them from exercising their criminal propensity. Zimmerman seemed to think it might be possible with a large-scale study to eventually predict a person's specific type of crime before that crime was ever committed.

ENTERPRISING DETECTIVES

Allan Pinkerton, born in Glasgow, Scotland, emigrated to the United States in 1842, and moved to Illinois to make barrels. His first engagement with law enforcement occurred four years later when he identified a gang of counterfeiters. Inspired by Vidocq, he soon became a deputy sheriff for Cook County and then the first police detective in the city of Chicago. He immediately solved a string of post office thefts and went undercover into the gang responsible for the crimes in order to identify its leader. His work on this case became news, which emboldened Pinkerton to found the Pinkerton National Detective Agency. He offered high-quality work and promised to accept no bribes, partner with local law enforcement agencies, refuse cases that initiated scandals, accept no reward money (his agents were paid well), and keep clients fully informed.

In 1856, he hired the country's first female detective, Kate Warne. As a childless widow, she'd answered his ad but initially he'd resisted, because the notion that a woman could be a detective seemed implausible. She assured him that there were important places a woman could go to get information, and methods she could use that were not accessible to men. When she persisted, Pinkerton gave her a chance and she proved her worth by showing how easily she could infiltrate high-class social events, where tips about professional crimes were gained with but a little flirtation. In addition, she assisted in uncovering plots against president-elect Abraham Lincoln, and accompanied him as part of the Pinkerton detectives escort to Washington; Warne posed as an invalid to stay close to him and watch for assassination attempts. She became one of the top agents in the agency, taking command of an all-female unit, but died young, with Pinkerton at her side. It was a sore loss for the man who'd initially been skeptical.

"The Eye," as Pinkerton came to be known, developed a reputation for solid integrity and delivery of services. His company motto was, "We never sleep," and he eagerly set up undercover operations, espionage for the government, and an interstate cleanup operation of the West's notorious bandits and gangs. Unfortunately, the Chicago fire of 1871 destroyed the impressive Pinkerton archive of documents and daguerreotypes, and the great detective himself died in 1884, although the agency continued his legacy.

In France, affairs among the famous brought attention to the microscope. The duc de Choiseul-Praslin, Charles-Louis Theobald, had married the daughter of one of Napoleon's generals. Her name was Fanny, and she bore him nine children. They hired Henriette Deluzy as a governess and the duke began an affair with her. Fanny fired her but Charles-Louis publicly continued seeing her, so Fanny announced she would seek a divorce. Shortly afterward, on August 17, 1847, the servants heard a scream from Fanny's bedroom, amid the crashing of furniture.

They believed a burglary was in process, but her locked doors hindered them from entering. From outside in the garden, they could see Charles-Louis framed in the window, so they returned to the room, which was now open. Fanny had been beaten to death with a blunt instrument and her throat was slashed open. Charles ran in and acted as if he'd only just discovered his murdered wife that moment. Since the servants had been unable to get into the room until after they saw him there, his story clearly had holes.

M. Allard, Vidocq's successor to the head of the *Sûreté*, surmised at once that the duke was lying. There was no evidence of a burglary and a pistol found under the bed belonged to him. It was covered in blood, as if used to bludgeon the victim. Charles-Louis claimed that he had run into the room with it, but seeing Fanny already dead, had dropped the gun to hold her. When he saw he couldn't help her, he went back to his room to wash off the blood, and a blood trail attested to that. The problem was, did he return to his room after killing her or after discovering her? In his room was a blood-stained dagger and the severed blood-stained cord from his wife's bellpull, which the servants had heard when the victim first screamed for help.

Allard arrested Charles-Louis and then set about trying to disprove the story he told. For assistance, he invited pathologist Ambroise Tardieu to observe the scene and the body. Tardieu had gained eminence with his study of asphyxiation victims, noting the differences among those who hanged or were suffocated by strangulation, chest pressure, or smothering. He even called the spots of blood that formed under the heart during rapid suffocation "Tardieu spots," after himself.

He examined the pistol under a microscope and located a chestnut-colored hair on its butt (the victim's hair color) and skin fragments near the trigger guard. In addition, the wounds on the duchess's head matched the size of the pistol butt. This evidence undermined the duke's story. Tardieu's reconstruction was that Charles-Louis entered his wife's room to slit her throat, but failing this, bludgeoned her to death as she screamed and fought. Her struggle seemed affirmed by a fresh bite mark on the duke's leg, which seemed to match his late wife's teeth.

However, Charles-Louis apparently realized that his story wouldn't hold up and he poisoned himself with arsenic.

MID-CENTURY PROGRESS

Trial judges were showing a more tolerant attitude toward physicians and other scientists. Throughout America, societies of medical jurisprudence were watching European developments and pressing for higher standards in their courtroom testimony. A trial that gained a great deal of notoriety involved physicians, anatomy experts, and dentists on both sides.

Among Boston's wealthy Brahmins in 1849 was George Parkman, near sixty years old. On Friday, November 23, he went out to collect his rents and never returned. He was last seen that afternoon at Harvard Medical College, where he had gone to call on Professor John Webster, who owed him money. Webster later claimed that he'd paid it and Parkman had left, but no one came forward to say they had seen him leave the building. Police searched the place, but found nothing to incriminate Webster.

His uncharacteristic behavior during the subsequent week alerted the building's janitor, Ephraim Littlefield. He drilled through a wall into the pit for Webster's privy—the only place in the building not searched. Once he broke through, he spotted human remains: a pelvis, a dismembered thigh, and the lower part of a leg. In short order, the police arrested Webster and searched the lab again. In a tea chest, they discovered a human torso and a dismembered thigh, while in the furnace lay charred bone fragments and a jawbone with artificial teeth that had not burned.

With so little to work with, it was difficult to determine the identity of the victim. Mrs. Parkman identified the body as her husband's from markings near the penis and on the lower back, and Parkman's brother-in-law confirmed that it was Parkman via the hirsute torso. Subsequent searches in the lab and office produced bloody clothing belonging to Webster.

Since they were already at a medical college with good facilities for the examination of a body, the investigators laid out the various body parts, tested them, and wrote thorough descriptions. By the end of the day, they had estimated the man's height to have been five feet ten inches—a match to George Parkman. The inquest jury pointed out that the two thighs found were different sizes, and the coroner patiently explained to these laypeople that one had been exposed to fire and the other was waterlogged from the privy. They could still be from the same person. The inquest and grand juries both ruled that John Webster should be tried for the murder of George Parkman. Judge Pliney Merrick and Edward Sohier defended Webster when the trial began on March 19, 1850, with Judge Lemuel Shaw presiding.

Attorney General John Clifford described how he believed Parkman had been killed, his skull fractured, and his various parts cut off and burned or dumped into a toilet. Dr. Jeffries Wyman, an anatomist, drew a life-size skeleton showing which parts of the body had been recovered and how they fit a frame the size of Parkman's. He said that in the furnace he had found bones from the head, neck, face, and feet, and he used actual bone fragments to demonstrate how they fit together.

Drs. Winslow Lewis and George Gay both helped to clarify the medical issues involved. Lewis, who was a former student of the defendant's, used Wyman's anatomical drawing to demonstrate the body's various parts, how they fit together, and how they had been affected by the attack. He said that there was an opening in the thorax region that might be a stab wound, but on cross-examination he admitted that he could not be certain.

By the third day, it was clear that the prosecution was relying heavily on scientific medical testimony, which was a boon for American medical jurisprudence. First, Dr. Oliver Wendell Holmes, dean of the Medical College, took the stand and said that someone with knowledge of human anatomy and dissection had done the dismembering. He also explained that a wound between the ribs would not necessarily produce a lot of blood, and that the remains were "not dissimilar" to Parkman's build—a careful sort of statement that would set a tradition followed even to the present day.

Then Dr. Nathan Keep, Parkman's dentist, insisted that the jawbone found in the furnace with the false teeth was in fact that of George Parkman. He recognized his own handiwork, and while the gold fillings had melted, there were "peculiar angles and points" on the teeth that he knew well. He had made a wax mold of the man's protruding jaw and filled it with plaster, and this he used to demonstrate how it matched the pieces of jawbone found in the furnace. Then he placed the loose teeth found in the furnace into his exhibit. It was a demonstration that initially impressed the jury, although dentists for the defense would try to undermine it.

Then came handwriting experts. Some witnesses talked about Webster's uncharacteristic behavior after Parkman disappeared, and finally, three letters were brought into evidence that had been written to deflect the investigation away from the Medical College, and all were unsigned. One had clearly been written by an educated man, but the other two diverged from that style. A man familiar with Webster's handwriting testified that he believed Webster had written all three letters. Even worse,

Webster's handwriting was recognized on the face of one of Parkman's loan notices, reading "PAID."

Sohier produced twenty-three character witnesses, and then brought out medical experts who insisted it was difficult to definitively identify these remains, or to say how the person had met his end. Dr. Willard Morton, a famous dentist around town, said there was nothing in the jawbone found in the furnace that would mark individual identification. Lots of people had protruding jaws. He produced a few false teeth of his own making that fit into the mold made by Dr. Keep. Once again, the jury members nodded, proving that it was a strong moment for the defense.

To round out the proceedings, Sohier said that the prosecution had to prove that the remains were those of Parkman, Parkman had been murdered, and Webster had not only done the deed but had exercised planning and malice. Clifford responded by reminding the jury of the strong medical testimony. He thought there could be no reasonable doubt that Parkman was dead and that he'd been located in pieces inside Webster's lab.

While this case marks a first for dental testimony, which presumably improved over the years, it's also notable for certain unusual procedures. Three of the doctors who had testified about medical facts for the prosecution also came in for the defense. Dr. Holmes mentioned the leading authorities on quantity of blood in a human body (not himself) and said that there was no way to tell with certainty whether a human bone had been broken before being burned. It was assumed that since medical expertise relied on objective knowledge and was therefore "neutral," it would not matter for which side they spoke, but their presence on both sides had the effect of undermining what these experts had said for the prosecution.

On the same evening that they went to deliberate, the jury had a verdict: Guilty. Webster was sentenced to be hanged. While he then admitted that he had murdered Parkman but had done so in self-defense, the sentence stood and he was hanged on August 30.

Another incident in 1850 opened up a new area for toxicology. In Bury, Belgium, Count Hippolyte de Bocarmé had married Lydie Fougnies, the daughter of a merchant, seven years earlier. However, her income was less than he had expected, so they knew their only hope lay in her father's fortune. It went to her brother, Gustave Fougnies, a sickly man, and as long as he remained a bachelor, they were optimistic that the money would come to them once he died. But then he decided to take a bride. So Hippolyte and Lydie invited Gustave to dinner. Lydie dispensed with the servants for the evening and served the food herself.

That very evening, November 20, her brother appeared to suffer from some sort of stroke. He fell over and died. Lydie and Hippolyte did what they could to save him, drenching him in vinegar. Then they cleaned up the floor and went to bed, leaving the body lying in a bedroom. The servants, who had witnessed this scene, sent for a priest, who in turn notified a magistrate. He noticed strange marks around the mouth of the corpse, and burns in the dead man's mouth and throat. He arrested the count and countess, pending an investigation, and ordered an autopsy. Gustave's organs were placed in jars and sent to the chemical researcher Jean Servois Stas (in some sources, Stass). He had studied in Paris under the likes of Orfila and now, at the age of thirty-seven, was a chemistry professor at the École Royale Militaire in Brussels. As a boy, he had set up his own chemistry lab in the attic of his parents' home, even making the instruments himself, and they were surprisingly precise.

While the victim appeared to have swallowed some kind of agent that burned, it was not clear exactly what this substance was. Given the use of vinegar (which he could smell), Stas believed Lydie and Hippolyte had attempted to neutralize or conceal some type of alkaloid, or vegetable-based poison. However, many chemists, including Orfila, were convinced that while metallic poisons could be identified, it was not so easy with alkaloids, which included morphine, nicotine, and strychnine. In fact, it seemed quite impossible, because tests used to detect poison destroyed the very substance being tested. But Stas was certain there was a way to isolate and test for any substance, so he worked on the problem for three months, dividing the victim's bodily tissues into several groups for different types of analysis, including smell and taste. Finally, using ether as a solvent, which he then evaporated to isolate the questioned substance, he found the potent drugs: coniine and nicotine.

A gardener recalled how Hippolyte had worked with tobacco leaves to create a perfume, and local pharmacists acknowledged that he had asked them questions about nicotine's toxic effect. In fact, he had practiced on animals and their corpses were available for toxicological analysis. (Stas also sacrificed several dogs to see if he got similar results.) Hippolyte's home was thoroughly searched and the authorities found a secret lab where he did the nicotine extraction. Stas even found traces of nicotine on the scrubbed floor. It became clear by logical deduction that Hippolyte had extracted this toxin from tobacco and force-fed it to the victim. When he used vinegar to mask the smell, it had caused the burns, and this gave Hippolyte away. With Gustave's already weak constitution, it had not been difficult, but even on a strong person, this poison could cause death within minutes, via respiratory arrest.

The couple stood trial together in 1851. With Stas's testimony, the killer was convicted, making Stas the first person to discover poisonous vegetable extracts in the body. Lydie begged for mercy and said that her husband had forced her into helping to kill her brother. The jury acquitted her, letting her go, while they sent Hippolyte to the guillotine.

Other toxicologists then developed qualitative tests with the Stas procedure to determine the presence of various alkaloids in the obtained extract. Yet with more work, a problem was evident: false reactions. At times, an alkaloid might form in a corpse that mimicked the test reactions for vegetable alkaloids. These substances were dubbed "cadaveric alkaloids" and they showed the need for better tests, especially in court. For every stride in forensic science, there was a dogged attorney to challenge it. That dialectic, while daunting at times, offered benefits to both sides. As new sciences emerged to play a part in the courtroom, these lessons were invaluable. The embarrassment of some professionals often lay the groundwork for others to refine the methods. Eventually the courts would set guidelines and standards, but for the moment, legal science was still experimenting with its role.

Throughout Europe, liberal reforms promised by successful revolutionaries failed to take shape and several royal rulers regained their footing, but not without having to take the people's representatives into account. Across Britain and Europe, the working classes formed unions to pressure for better wages and workplace conditions, while socialist creeds denounced private enterprise monopolies. Certain countries gained a strong and cohesive identity, and from among them different countries dominated for different reasons. Germany rose to political power in Europe just as the United States divided into North and South for a bloody four-year Civil War.

Yet science continued to prosper, including discoveries that would assist with better criminal investigation. The year 1851 has often been viewed as the real start of the age of science, when London put on a "Great Exposition of the Works of Industry of All Nations," building the astounding seventeen-acre Crystal Palace to showcase the achievements of the Industrial Revolution. All nations were invited and more than one hundred thousand exhibits were offered, launching the first in a series of World's Fairs that would pit nations against one another in a friendly rivalry for the largest and most awe-inspiring designs. With this kind of social and cultural encouragement, more innovations were on the way. It was considered the triumph of "the useful arts" over nature. One exhibit, the binaural stethoscope, established the principle of reading cardiac patterns, which would later influence the design of the first lie detectors. The hope for the exposition was that industrial aggression would replace military aggression, although in fact the former would feed the latter. The exhibit, and others that followed in other countries, celebrated human invention, and the crowds that arrived by rail affirmed the culture's interest in material progress.

EMPHASIS ON EVIDENCE

In several papers, Mathieu Orfila had listed the types of insects and arthropods that visited a decomposing corpse to feed and lay eggs, adding to the work done by others a century earlier. He demonstrated the correct succession of insects to a corpse, which aided in a case two years later. A couple had moved into an apartment in Paris and decided to have it renovated. To their horror, while repairing several ill-fitting bricks on a mantel, the workmen discovered the mummified corpse of an infant lodged between the walls. That placed the couple under suspicion.

In their defense, Dr. Marcel Bergeret, from the Hôpital Civil d'Arbois, reevaluated the situation. It had been his private interest to study the changes that cadavers underwent after burial, so he was something of an expert on insects. He found that a specific type of fly, Sarcophaga carnaria, which gravitated toward dead flesh, had deposited larvae on the dead baby, and he was able to determine that the child had been born alive and full-term, as early as 1848. The next year, mites had laid eggs on the dried remains, so he concluded that the baby had been placed in the walls before the current residents had even moved in. Suspicion then fell on the previous occupants, but over the previous three years there had been four separate tenants. An investigation ensued, eliciting information from the landlady that a young woman who had lived there had appeared to be pregnant, but this landlady had never seen a child. The suspect was located and arrested, but when the infant's remains failed to yield a cause of death, the woman escaped conviction. Bergeret published a description of the case, adding that there was an urgent need for more study of forensic entomology.

This case inspired widespread interest among pathologists, notably Edmond Perrier Mégnin in France, who worked at the Natural History Museum. He regularly visited morgues and cemeteries, recording eight distinct stages of necrophilous insect infestation. He agreed to consult on a similar case in 1878 of a mummified infant, utilizing the information that was then known about insects on corpses, as well as knowledge about decaying plants, to estimate that the child had been dumped some seven months earlier. He would eventually publish *Fauna of the Tombs* in 1887 and *Fauna of the Cadavers* in 1894, which offered his considerable findings about both insects and plants. He identified those insects that assisted in estimating the postmortem interval over the course of three years, should a corpse remain out in the open that long: egg-laying blowflies, beetles, mites carried by beetles, moths, and flies that liked fermented protein. The progression was different for buried corpses, and Mégnin wisely warned that the results might differ in other types of soils and climates. Other researchers during this time joined him by studying the corpses of animals or going to exhumations to look at the effects of burial on the dead. Their work helped to identify the bits of insects that police often mistook as the marks of weapons or poisons. Eventually researchers across the ocean in the United States and Canada would contribute to the growing body of data. By 1897, M. G. Motter, who had assisted with the disinterment of a cemetery in Washington, D.C., would identify different types of insects inside the coffins.

DIVIDING LINE

Back in America in 1852, an Oregon sheriff was asked for his opinion about the nature of a hole found in a victim's shirt. He used the suspect's firearm to shoot another hole into the material to prove that the original was not a tear but identical to the place where the test bullet had entered.

Then the 1853 trial of John Hendrickson in Albany,

New York, caught America's attention. Hendrickson, twenty, was suspected of killing his wife, Maria, who was nineteen, because their two-year marriage had been rocky and Hendrickson had assaulted another young woman. In addition, the baby that his wife had borne had died in their bed the year before. The examining doctors found symptoms that indicated that Maria may have been poisoned by aconitine, via aconite. Yet Maria had used homeopathic remedies with this ingredient, and she'd been feeling ill in the days prior to her death so it would be difficult to prove that she'd been deliberately poisoned, especially with the state of toxicology at the time. While Stas had demonstrated an ability to find alkaloids in human tissue, his experiments had been arduous and no one in the States yet had repeated them, especially for an entirely new substance—and one considered the most difficult to detect.

Hendrickson was indicted for murder, because his poor behavior around town garnered circumstantial evidence against him, as did the rumor that Maria, whose family was wealthy, had intended to divorce him.

It soon became evident that there was no way to prove that aconitine was in Maria's system, so Hendrickson's defense attorney attacked the medical diagnosis of poisoning. Dr. James Salisbury, an expert medical chemist, had conducted rudimentary chemical tests that he believed confirmed the presence of the poison, as well as smelling and tasting the substance, but he had failed to save the aconitine that he'd isolated. Indeed, he'd fed it to a cat, and the cat had survived. The defense attorney seized this with a flourish: "The cat should have died out of deference to the doctor's opinion, or the doctor should have given up his opinion out of deference to the life of the cat." Salisbury was embarrassed by the outcome, but he believed in his testimony.

The prosecutor summarized the case as perhaps the most important one of the century, in light of the medical and chemical questions it raised. His closing remarks emphasized the experiments that Salisbury had performed, which shored up his testimony. He used the circumstantial evidence as reinforcement. The jury found Hendrickson guilty, and medical science proclaimed this a victory.

However, Dr. David Wells read about the way Salisbury had claimed to isolate aconitine and he knew that it was impossible, so he made a public rebuttal of what he considered a wrongful conviction. He insisted that Salisbury had derived a different substance that might taste like aconitine, but he certainly had not proven that he'd gotten that alkaloid. Wells wrote to the leading medical men and toxicologists, urging them to join in his protest. Many responded, and a joint letter was drafted to the governor of New York. They believed that the jury had been swayed by the confident manner of the expert witnesses, not by their so-called proof. Even as Hendrickson waited to be hanged, scientists across the nation rallied on his behalf, discussing the case at medical meetings. But the New York Supreme Court and the New York Court of Appeals both affirmed the verdict. Hendrickson's new execution date was set, and despite a flurry of medical protest, he was hanged on May 5, 1854. Between two thousand and three thousand people waited outside the Albany prison to try to see, including a large contingent of the press. For the professionals involved, the case proved traumatic and they renewed their efforts to introduce better standards and safeguards into their profession. That a man could be executed on such flimsy and controversial evidence seemed ghastly to them.

In England, another costly error was made by Alfred Swaine Taylor, who had studied with the famous Orfila and become professor of medical jurisprudence at Guy's Hospital Medical School. In 1836, he'd published *Principles and Practice of Medical Jurisprudence* and would eventually become one of the century's most eminent forensic scientists, although his early cases were setbacks for toxicology. Among them was a notorious serial poisoner.

In 1855, Dr. William Palmer, a gambler deep in debt, was accused of using strychnine to poison a friend named Cook, who had died and "left" him a considerable sum. He was allowed to attend the autopsy and was caught trying to abscond with a jar filled with the contents from the dead man's stomach. Palmer was already associated with a string of sudden and unexpected deaths among friends, acquaintances, associates, and relatives, including his rich wife (whom he had insured for a grand sum) and his mother-in-law. A few of these people had died quite suddenly while staying with Palmer. He'd even insured his own brother, Walter, and invited him into a bout of drinking, whereupon Walter expired. The insurance company refused to pay, so Palmer insured someone else, who also quickly died. Suspicious investigators decided to exhume a few of these victims and they found a large dose of antimony in Palmer's deceased wife.

Taylor analyzed the contents of Cook's stomach, and while he found no strychnine, he did detect some antimony. Still, it was not in a sufficient quantity to kill a man. No one thought that Palmer's possession of the stomach contents for a short time might have influenced the results, and Taylor's testimony at the trial was unconvincing. The circumstantial evidence did Palmer in and he was hanged in 1856. The rope used was cut into pieces and sold for a good profit. Taylor had tried his best, but some poisons escaped the body in the form of gas. He simply could not prove the presence of something that was not there. His next trial, too, proved problematic, not just for him but the entire profession. In America, several toxicologists who had paid attention to the Palmer trial set themselves to the challenge of detecting strychnine. T. G. Wormley and John J. Reese both committed themselves to research and by the end of the decade were able to publish papers that contributed to future solutions. Clearly, toxicologists were determined to improve their examinations and proofs.

But in the meantime, other scientists were making discoveries applicable to the forensic arena. Between 1853 and 1856, Ludwig Teichmann of Poland invented the first microscope crystal test for hemoglobin, and Richard L. Maddox developed dry plate photography, which reduced the exposure time and amount of equipment formerly required for processing. This made his technique highly practical for photographing inmates. Prisons collected a catalogue of images to help them keep track of convicts who were repeat offenders. Sir William Herschel, a British officer, began to use thumbprints to authenticate documents and he learned from his own studies that fingerprints did not change with age.

In 1859, the United States became the first country in which photographs were used as evidence in a court of law, while in Germany, physicist Gustav Kirchhoff and chemist Robert Bunsen founded the field of spectroscopy with a prism-based device that made it possible to study the spectral signature of chemical elements in gaseous form. Bunsen had met Kirchhoff at the University of Breslau, where Kirchhoff taught. When Bunsen took a position at the University of Heidelberg, he secured one there for Kirchhoff as well so that they could continue to work together. Bunsen had developed techniques for separating and measuring chemical substances, and he invented the Bunsen battery and the Bunsen burner. The latter offered a nonluminous flame test for metals, and this paved the way for the spectroscope, which demonstrated that the color of a flame can be used to identify the substance burning by separating light into component wavelengths. These two scientists discovered that the spectrum of every organic element has a uniqueness to its constituent parts. By passing light through a substance to produce a spectrum, the analyst could read the resulting lines, called "absorption lines." That is, the specific wavelengths that are selectively absorbed into the substance are characteristic of its component molecules. Then a spectrophotometer measures the light intensities, which yields a way to identify different types of substances. While not immediately relevant to the courts, it would eventually become an integral part of chemical analysis for trace evidence.

Over the next few years, presumptive tests were also developed for detecting blood on smooth surfaces or clothing, although not yet distinguishing human blood, and German pathologist Rudolph Virchow systematically examined the value of hair as evidence in crime detection. He successfully persuaded investigators to include his ideas in their repertoire. Each of these procedures advanced forensic science and the more techniques there were that proved reliable, the more it became evident that crime investigation must integrate them.

Charles Darwin had published *On the Origin of Species by Means of Natural Selection* in 1859, in which he proposed a theory of evolution based on survival of the fittest. Realism in art and literature helped to reinforce the new scientific ideas, and evolutionary theory soon provided a perspective for developing theories of criminology.

That same year, Dr. Alfred Taylor experienced his next humiliation in the courtroom. Doctor Thomas Smethurst was charged with the murder by poisoning of Isabella Bankes, a woman he had married bigamously. When she died, he proved to be the beneficiary of nearly two thousand pounds, and two physicians had stated that the decedent's symptoms had resembled poisoning. More suspicious, his other wife took him back, inspiring the notion that she'd known he'd been with Bankes merely as a means to acquire her money. Taylor was called into the case.

He indicated that he had found traces of arsenic in the body and in a bottle of colorless liquid found in Smethurst's rooms, but during the trial when the bottle of liquid was proven to be potassium chloride, Taylor admitted that his findings were the result of an imperfection in the apparatus that he'd used. He'd actually detected potassium chlorate. Furthermore, he and two other medical experts contradicted one another on matters in which there should have been no dispute, leaving the case in a muddle, but Smethurst was nevertheless convicted and sentenced to hang.

The press jumped on this case, as did other professionals, criticizing the expert testimony and asking how such eminent people could disagree on matters based in proper scientific analysis. A collection of medical men composed and signed a letter to the home secretary insisting that the case receive further attention, as there had been no clear proof of poisoning. Officials sent it to yet another medical expert for analysis, and finally the defendant won a pardon. The way his case had been handled brought shame to the profession and detractors of forensic medicine dubbed it a "beastly science," which forced physicians to greater accountability. The toxicologists and anatomists had to work hard to recover their dignity and credibility. But eventually, they did.

FOUR

OUTER MAN, INNER MAN

THE MEASURE OF A MAN

As cities expanded during the nineteenth century, crime rates rose, and it was no longer sufficient to be able to identify a thief or burglar on sight. Few investigators possessed the prodigious memory of Vidocq for case details and criminal faces, and while photography was used in some places as a means of preserving images, the number of photographs added up and took up space with little sense of organization. No one had yet devised a system for sorting through them, and since criminals often used disguises, pinning crimes on specific perpetrators often required specific identifiers. People involved in law enforcement and science were working on that goal.

Christian Freidrich Schonbein developed the first presumptive blood tests in 1863, basing his discovery on observation that the reaction between hydrogen peroxide and hemoglobin took on the appearance of "foaming" as the oxygen bubbles rose. Schonbein reasoned that if an unknown stain foamed when hydrogen peroxide was applied to it, then that stain probably contained hemoglobin, and therefore was likely to be blood. However, he could not prove that it was human blood, specifically.

Other scientists studied the structure of the body. Belgian statistician Lambert Quételet set forth the notion in the early 1860s that no two human beings shared the exact same dimensions. Over two decades before, he had published *On Man*, in which he proposed statistical studies of intellectual development. The warden of Louvain Prison agreed with Quételet about the physical dimensions and applied the hypothesis to his charges by taking their physical measurements. Yet while Quételet's methods failed to catch on with a wider audience, physical anthropology, which had been in the lexicon since 1593, dominated mid-nineteenth-century discussions about crime.

Among those who made a significant contribution was a man interested in human remains. In Paris, the ancient cemetery of the Celestins was undergoing excavation, and while many of the remains had decomposed into dust or bone fragments, some skeletons remained intact. Paul Pierre Broca, a former child prodigy with academic degrees in literature, mathematics, medicine, and physics, was a professor of surgical pathology at the University of Paris. He was interested in cartilage and bone structure, as well as cancer, the behavior of blood, and the mechanisms of the brain in language processing. He would soon gain the distinction of having his name applied to the part of the frontal lobe of the brain, Broca's area, which he proved, via intensive study of aphasic patients, was responsible for speech production.

It was his work in the cemetery in 1847, however, that led him in another direction. Broca was well-read in anthropology and although he was denounced as a materialist corruptor of youth for his support of Charles Darwin's theory, in 1859 he founded the Anthropological Society, along with a laboratory at the École des Hautes Études and the Anthropological Institute in Paris. The Roman Catholic Church stood firmly against the Anthropological Institute, since it encouraged study in the mutability of races and species, and Church officials were incensed when Broca produced a journal, *Revue d' Anthropologie.*

While attempting to measure the skulls of the deceased, Broca invented more than two dozen new measuring devices called craniometers, which standardized a classification system known as craniotomy, derived from the published accounts of the anatomical measurements of skulls in humans and animals. In part, this work had been done to distinguish Europeans from other races via skull capacity and brain size, as well as to illustrate human kinship with apes. At the Anthropological Institute, Broca and others taught anthropometric measuring in the hope of learning more about human intellectual capacity, which they expected would help them to understand the differences among the various races. They were hoping to prove superiority for males over females and

European races over others, and Broca actually warned against mixing Africans and Europeans. These freethinkers believed that most people could be improved to European levels, but not all. Some races supposedly suffered from an "organic curse."

Phrenology had already won adherents to the idea that specific brain functions and personality temperaments showed up in bumps and depressions on a person's skull, and this idea won new adherents during the 1860s. Some "practical phrenologists" who claimed an elevated position with respect to knowledge about people would "read heads" for fees, as well as put on demonstrations, which helped to inspire the anthropological enthusiasm for measuring heads. They would run their fingertips over the skulls to identify the precise location of elevations and depressions, and compare their findings to charts (or ceramic busts) that showed how the brain areas were divided. Thus they could diagnose a particular temperament and, in criminal cases, predict the potential for reform. Some physicians decided that criminals could be detected via facial characteristics, so for a time, prisoners were classified by these features. While the phrenologists turned out to be partially correct about the idea that different parts of the brain had localized functions, their deductions about skull formation and personality diagnosis were uncorroborated with actual science.

In addition to criminality, the cause of insanity continued to interest investigators and some began to look at the notion of heredity. Several sociologists studied family lines to prove that some genetic strains produced outstanding citizens and others were only good for spawning losers. A specific type of criminal, who showed extreme brutality and no remorse, got attention as well. To classify such people, Philippe Pinel had introduced the notion of "mania without delirium" in 1809, and more than two decades later British physician James Prichard called the same behavior "moral insanity," to indicate that one's faculty for appropriate behavior had been corrupted. These were the precursor labels to the psychopath—the person who had no conscience about his cruelty or destructive acts. Psychiatrists continued to try to learn what they could about the condition, while anthropologists were convinced they could formulate identifiable physical characteristics.

Among these ruthless and remorseless criminals was Edmond de la Pommerais, who in the age of evolution was attempting his own form of survival of the fittest. He had purchased a practice in homeopathy but had failed to see the riches he'd expected, so he looked around for an easier route to wealth. To facilitate his introduction to women of means, he called himself a count. Soon he married a woman from a wealthy family whose mother kept strict control over her, but when the mother fell ill and died, de la Pommerais used the inherited funds to purchase a lavish lifestyle. Soon he and his wife were bankrupt. He insured his mistress, Séraphine de Pauw, for a considerable sum, and she, too, fell ill and died, which filled his pocket for a little while. Then he pressed de Pauw's sister into a scheme for insurance fraud, which included making out a will in his favor. He assured her that if she went along with it they would both benefit substantially.

The police got wind of this plot from an anonymous letter and were able to save a potential victim by stopping her from carrying it out. Having discovered what de la Pommerais was up to, they then exhumed Séraphine and found that she had not died of cholera, as de la Pommerais had indicated. Professor Auguste Ambroise Tardieu, who taught forensic medicine at the University of Paris and was considered an expert on the wounds and asphyxiation symptoms of victims of hanging, supervised tests on Séraphine's body for arsenic and antimony, but these were negative. Still certain that the victim had been poisoned, Tardieu decided to perform an experiment. He remembered that before Madame de Pauw died, she had suffered from a racing heart, so he suspected an alkaloid toxin. In order to prove this, he injected a dog (some sources indicate it was frogs) with an extract he had obtained from the decedent's organs using the Stas method. The animal vomited and showed symptoms of a racing heart. Then Tardieu found the drug digitalis among de la Pommerais's remedies, which was used for regulating the heart but could be lethal in elevated doses. But Tardieu needed more than just the presence of the drug in the "count's" stock. When the police managed to scrape up traces of vomit from the victim's sickroom, Tardieu tested it for digitalis and proved his theory. This evidence held up in court and on June 9, 1864, de la Pommerais was convicted of the murder of his mistress. He was executed.

That same year, Henry Goddard, formerly a Bow Street Runner and now a private detective in London, investigated another insurance swindle. The Gresham Life Assurance Company requested his assistance in investigating the unexpected death of one of its employees, Edward J. Farren. Goddard accepted the case and learned that Farren had been traveling when he'd died and that his wife had received a letter about his death from strangers, but there were inconsistencies in the stories. Goddard conducted the investigation with his characteristic thoroughness. First he learned everything he could about the victim, including that the man had a deformity of the foot that made one heel three inches higher than the other one, resulting in a noticeable limp. Thus, if the man had not died but was attempting to collect money from a faked death, he would have had to hide the limp. That is, he'd need a cobbler who could create a special kind of footwear.

Goddard asked around and eventually found the tradesman who had made a special boot for a princely sum. But reportedly the man who had collected the boot still limped slightly, which gave Goddard sufficient detail to question concierges around the city about seeing such a person in their hotels. With some extraordinary sleuthing, Goddard finally traced the still-living Farren as far south as Egypt and all the way to Australia. In a hotel there, he located the specially made boot, along with its owner, clinching the case for the "assurance" company, which paid *him* instead of Farren's "widow."

MAKING A MATCH

A case from the history of ballistics actually belongs with document examination. In 1860, a man was found in a London street, shot dead. Beside him lay a crumpled piece of newspaper from the London *Times* dated March 24, 1854. Since the paper smelled of gunpowder, the piece was collected as potential evidence. When a suspect was identified by the name of Richardson, the police searched his rooms and found a double-barreled pistol. One barrel had been fired but the other was still loaded; it had newspaper wadding stuffed into it, also from the *Times*. Although it bore no date, the investigator went to the newspaper editor and asked for back copies for comparison. Both pieces proved to be from the same issue. Under pressure from this evidence, the suspect admitted to the murder.

Ballistic interpretation managed to get into the American Civil War, when a general for the South, "Stonewall" Jackson, was shot while riding in the front lines. The bullet removed from him proved to have come from a gun fired by one of his own men, and thus he succumbed to friendly fire. His death was a significant loss for the South, coming just before the turning point of the battle in Gettysburg, Pennsylvania, in 1863.

In 1869, Swiss biologist Frederick Miescher carried out chemical studies on the nuclei of pus cells from discarded bandages that isolated DNA for the first time. He called it "nuclein" and demonstrated that it consisted of an acidic portion and a basic protein portion. The acid portion is now known as DNA, while the rest is responsible for DNA's packaging. But it would be more than a century before this discovery proved useful to investigators. For now, they had to rely on ingenious ideas; several occurred in cases later that year, just after Paris established the Institute of Legal Medicine.

The *curé* of Bretigny was the victim of a fatal shooting, but the bullet fragments removed from his head during the autopsy proved too fragmented to match to a gun or to bullets owned by the primary suspect, a watchmaker named Cadet. Not to be stymied, French chemist Monsieur Roussin melted the fragments to establish their physical composition via chemical analysis, and they proved to be a compositional match to bullets found in Cadet's residence. Based on this evidence and reports of bad blood between the two men, Cadet was convicted of murder.

Also in Paris, an incident occurred in the history of serological analysis that demonstrated the power of science and deductive reasoning. Water from a well in the cellar of a restaurant on rue Princesse had made several customers ill, and an investigation yielded a package that contained the lower half of a decomposing human leg. A new officer in the French *Sûreté*, Gustave Macé, also an attorney, was assigned to look into the matter. When he peered into the well, he spotted another package, so he drew it out and opened it to find the lower part of yet another leg, encased in a stocking and presumably from the same victim. A physician determined that the remains were from a female victim.

Both legs had been stitched into a black calico bag, tied with a method known among tailors, so Macé kept this information in the back of his mind. Looking through the city's files of missing women, Macé was not certain which lead to follow, but Dr. Ambroise Tardieu offered his own opinion that the legs belonged to a man, not a woman, who had probably been dead at least a month and a half. So Macé started over again, searching files of missing men. He remembered the human thigh that had turned up around that time as well as a thigh bone, both found in another area of town. Indeed, chunks of human flesh had also been discovered in the St. Martin canal, and a man had been spotted dumping them "for the fish," but unfortunately he had not been detained. The gendarmes had also stopped someone early one morning carrying several leaking parcels, which he claimed were hams, so they had not bothered to open them. The description they gave matched that of the man who had tossed the remains into the canal, so as aggravating as their lax attitudes were, these officers did provide some valuable pieces to the puzzle. Pulling all the items together, Macé believed that the two legs he had at the morgue belonged to those earlier pieces.

Through intensive detective work and a lot of questions asked, the victim's identity was finally revealed: a recluse named Desiré Bodasse. Macé also learned that a tailor named Pierre Voirbo had visited the household that used the well, bringing work to a woman in an apartment upstairs. He then determined that Voirbo had known the murdered man and during a search he found some of the victim's items among Voirbo's possessions. Macé interrogated Voirbo to learn more about their association. It turned out that recently they had quarreled over money and that Voirbo had cashed in some of Bodasse's Italian stocks. He had also collected newspaper articles about the mystery of the legs found in the well. Macé noticed that Voirbo's former lodgings had been cleaned (and the cleaning woman said that Voirbo had done this himself), which meant that evidence was destroyed, but then he realized that the floors were tiled, with alleys between the tiles. Having a flash of inspiration, he poured water on the floor to see where it ran. He then lifted the tiles in that area, under a bed. Beneath them was enough blood to indicate that something of a violent nature had occurred in that room. Voirbo, who was forced to observe this innovative demonstration, broke down and confessed to bludgeoning Bodasse and carving up his body. He blamed his father, who had often threatened him, for his impulse to commit murder. Before Voirbo went to trial, he committed suicide with a razor smuggled into the prison.

Macé had demonstrated how a systematic, logical approach to investigating crime got results, and he grew famous for the way his analytical thinking brought murderers to justice against all odds. He was one of the investigators featured in papers that contributed to the popularity of the brilliant mind as a detective's trusty tool. Macé went on to become head of the detective division, and among his practices was to require photographs of all criminals.

Odontology came into a case in America when a woman was found dead in Ohio in 1870, with five distinct bite-mark bruises on her arm. Since she had been the mistress to a man named A. I. Robinson, he became a suspect, but there was no evidence against him aside from innuendo, and there was also another suspect. An examining dentist who hoped to identify the culprit via matching the bruises on the body to Robinson's teeth actually bit the dead woman's arm himself to show the difference between the mark he'd left and the suspect's. Then he had Robinson bite *his* arm for making the comparison. Another dentist, Dr. Jonathan Taft, cast Robinson's teeth, using this to show the spot where one was missing, which made his mouth fit perfectly into the wound pattern.

However, the defense attorney raised issues about how similar teeth are and how difficult it is to tell what a skin bruise means, so Robinson was acquitted.

Yet forensic botany got a boost in 1873 when pathologist Alfred Swaine Taylor examined microscopic plant life in the lungs of a drowning victim to indicate that the person had drowned in a body of water apart from the one in which the corpse had been found. Thus, it was a case of murder, proven with diatoms.

The following year, London investigations advanced. The Prisoners Property Act of 1869 had given authority for police to retain certain items of prisoners' property for instructional purposes, and by 1874, the Central Prisoners Property Store had enough items at Scotland Yard for an Inspector Neame to devise the idea of a crime museum for the purpose of giving practical instruction to police officers. It opened in 1875 and two years later, thanks to a reporter barred from entry, acquired the name "The Black Museum." It later moved with the Metropolitan Police Office to New Scotland Yard.

CRIMINAL TYPES

In 1876, members of the Society of Mutual Autopsy, many of them the leading anthropologists of Paris, pledged to donate their brains after death to science. Since they regarded themselves as members of an elite race, they were confident that a study of their brains would advance the science of man significantly, particularly in light of Broca's discovery that a specific part of the brain was responsible for such a sophisticated activity as speech. Once again, they defied the church's teachings about the sanctity of the body. In fact, many did not believe in the soul's immortality and they viewed the body after death as valuable only for scientific study.

The freethinking anthropologists gained sufficient political power in France to remove some of the religious iconography from public works, and they renamed a few streets. They hoped for the gradual evolution of other cultures to their level. Among the members was a statistician and demographer named Louis-Adolphe Bertillon, who had been quite excited by the earlier work of Quételet. He believed in "social physics," which hypothesized a way to use numbers and calculations to understand human beings. He was among the thinkers to establish the School of Anthropology, and his two sons, Jacques and Alphonse, would follow in his footsteps. While Jacques became a statistician as well, Alphonse took the ideas much further. He set to work on creating an interesting invention, the *portrait parlé*, which was a full description of a criminal, including hundreds of pieces of information from eye color to scars to a person's posture. The task required a good eye and a great deal of patience, but it would not really be used until after the young, introverted Bertillon had established himself in another way.

Before that happened, another man was already stealing the show. Even as the anthropologists in Paris were pledging their brains, Italian anthropologist Cesare Lombroso, a former army surgeon and a professor at the University of Turin in Italy who owned the world's largest collection of skulls, published *L'uomo delinquente*. Experienced in the methods of phrenology and influenced by Broca's ideas about racial types, Lombroso believed that human behavior could be classified via objective tests. He had made numerous measurements of both lunatics and criminal offenders, dead and alive, and studied even more photographs of them.

Lombroso's data convinced him that there were anatomic differences between normal people, lunatics, and criminals. That is, certain people were born criminals and they were identifiable by specific physical traits, such as bulging brows, close-set eyes, protruding jaws, disproportionately long arms, and apelike noses. In other words, delinquency was a physiological abnormality that could be observed in someone's simian appearance. As well, they tended to acquire tattoos, speak in slang, lack foresight, act impulsively, experience violent emotions, and prefer idleness and play.

He believed there was such a thing as a "born criminal" who was irresistibly compelled toward a life of crime, and that this criminal was an atavistic being—a throwback to earlier hedonistic races. Others could become criminals through weak natures or "vicious training." The born criminal had peculiar sensory responses, a diminished sensibility to pain, no sense of right and wrong, and no remorse. Also, it seemed, only criminals bore tattoos, which Lombroso considered a reversion to ancient tribal rites, primitive races, and the craving to torture, mutilate, and kill. It was suggested by those who reported this work that the police could make arrests more accurately if they trained themselves to spot the right traits. And the public could better protect itself from a stranger who had an obvious criminal appearance.

Striking a social chord, Lombroso's ideas spread across Europe and America, supported by the new evolutionary thinking. The theory of the born criminal inspired widespread prejudices that made those without these traits feel smugly superior, and it often victimized innocent people. But the theory dug in and remained a compelling idea for several decades.

Yet in 1875, sociologist Richard L. Dugdale announced in the annual report of the Prison Association of New York that he was making a study of an American family whom he called the "Jukes." Dugdale had inspected thirteen county jails around New York State and asked a number of prisoners the same set of questions about the environment and families in which they had been raised. In one county, he located six persons who were blood relations and who were all being held on some criminal charge. He discovered that this family had a long lineage, so he set about to study them in the interest of seeing if heredity, coupled with a certain environment, influenced—perhaps even caused—criminality.

Dugdale's report was published as a book in 1877. He claimed to have examined several generations of this family, and out of some 540 descendents of six original sisters, the Jukes showed a higher than average percentage of syphilitics, prostitutes, thieves, and murderers (about 140 offenders). This degenerate tribe was compared against another family of good Puritan stock, out of which had emerged mostly upstanding citizens, and even some presidents. While it seemed a good argument for heredity, and Lombroso was an enthusiastic supporter of this work, Dugdale indicated, contrary to Lombroso, that environment could be a factor as well (although Lombroso later accepted this idea). Eventually, Dugdale's work was discredited, since the Jukes were not all from the same family.

WHAT THE BODY REVEALS

While science was improving methods of investigation and prosecution, as well as trying to advance the understanding of criminal behavior, it was clear that a more definitive approach was needed to process and identify repeat criminals—and there were many, since police were only just appearing in major cities. The science of identification took two directions. Some men continued to focus on human anatomy as a whole and others on only a specific area of the body: fingerprints. The history of identification methods can thus be likened to several independent vines intertwining, as these men crossed one another's paths, sometimes helping, sometimes thwarting others who were making similar discoveries.

In 1877 in the United States (the year that the Commonwealth of Massachusetts replaced its coroner system with a medical examiner), Thomas Taylor proposed that skin-ridge patterns on palms and fingertips be used for identification, although he did not take it any further. That same year William Herschel, a former civil servant in India, brought fingerprinting methods to the attention of the inspector-general of the Bengal prison system. Herschel had already used fingerprints to seal contracts and then, as a magistrate, had identified men attempting to collect their pensions twice. The official ignored him.

In 1880, Scottish physician Henry Faulds, who had developed a way to make "fingermarks" visible with powders, thanks to perspiration, successfully eliminated a suspect and helped to convict the true offender in a burglary. He had already written about fingerprints left on Japanese pottery and while he was developing a thesis about racial identification, the crime occurred. The thief left sooty prints on a white wall, making the ridge patterns nicely delineated. Faulds made a match to the thief and even managed to intimidate him into a confession. He then published "On the Skin-Furrows of the Hand" in the prominent scientific journal *Nature*.

Yet there was little movement initially to get widespread acceptance of this method. Faulds had penned a letter to the editor of *Nature* about his ideas, which was published. Herschel read it and responded with a letter of his own, claiming to have used fingerprints for identification as far back as 1858. Apparently this enraged Faulds, a jealous and competitive man, so he traveled to England to assert himself as the discoverer. He acquired a position as a police surgeon and then importuned Scotland Yard to take him seriously. But his personality got in the way; he became bitter and angry when things failed to go smoothly. Thus, no one was willing to further his claims or open doors.

One man who attempted to reach Faulds was Francis Galton, cousin to Charles Darwin, who was enamored of the statistical approach to identifying humans. He had made comparisons between body measurements and sensory acuteness, and he had noticed the differences in fingerprints. He was working to make his own contribution to the enterprise.

Fingerprinting methods were clearly in the air, because in 1883, Mark Twain published *Life on the Mississippi.* In one chapter of this novel, a man describes how he relied on a bloodstained thumbprint to track down the murderer of his wife and child. He mentions that a prison warden had told him that thumbprints never change and that no two men have the same patterns. The man uses the print to identify two culprits, one of whom he executes. However, before the system of fingerprints would come into its own, another development detoured criminal identification on to a different track, and this approach, too, relied on the scientific method.

Alphonse Bertillon, a petulant file clerk for the French police and the son of the anthropologist Louis-Adolphe Bertillon mentioned earlier, had grown frustrated during the late 1870s over the enormous and chaotic collection of photos in the police bureau. As well, he had to deal with more than five million imprecise files, thanks largely to the system that Vidocq had devised half a century earlier. He had read Quételet's publication *Anthropometry, or the Measurement of Different Faculties in Man,* published in 1871, and he firmly believed that he could develop a systematic way to measure and categorize the criminals who were arrested. Like the men who'd been colleagues of his father, the younger Bertillon believed that human measurements fell into specific statistical groupings, but there would still be defining differences from one individual to another. He calculated that if fourteen different measurements of a person's body were taken, the odds of finding two with the exact same measurements of every feature were 286,435,456 to one.

Bertillon decided to experiment. For each person arrested, he took between eleven and fourteen key measurements, from the length of the foot to the width of the jaw, from the width and length of the head to the length of the forearm, classifying each person and recording the measurements on cards. His system derived from another assumption derived from anthropology, that adult human bone structure does not change, and it involved three steps: the body measurements, which had to be taken under controlled conditions in a precise manner; a description of the body's shape and movements; and a description of identifying marks, such as moles, tattoos, deformed limbs, or scars.

He wrote out his ideas and submitted them to the prefect Louis Andrieux, who basically ignored him. When Bertillon persisted, Andrieux told him to desist and even wrote a letter to the young clerk's father to complain that his son was annoying and was possibly even quite mad. Louis-Alphonse, anxious over Alphonse's employment opportunities, asked to read the report himself. He was prepared to insist that his son behave more responsibly, but when he read what Alphonse had written, he was impressed. He believed that this system could work, not only in this prison, but might extend well beyond organizing files to become a true phenomenon. Alphonse, he saw, had managed to apply what he and his colleagues had spent their lifetimes trying to prove. He wanted this idea to get the attention it deserved.

Still, the older man could not persuade the prefect to consider it. Alphonse suffered the scorn of his colleagues for his failed attempt to incorporate changes, and he stewed in his anger. He set about writing his own book, which he would call *The Savage Races,* while awaiting a changing of the guard. It happened, and Jean Camecasse became Bertillon's immediate superior. Louis-Alphonse brought his influence to bear, along with that of an attorney, Edgar Demange, who was sympathetic to the young clerk's notion. In November 1882, Camecasse made a momentous proposal. He told Alphonse that if he could prove his system in three months by identifying a repeat offender, Camecasse would consider adopting it.

That did not provide much time for a scientist, and it required that a recidivist be arrested during that time frame, but Alphonse believed he could manage it. He had to. He began right away, keeping careful records of each man he measured and adding a full-face photograph and a side profile. He included any identifying marks, such as a scar or missing tooth, and made notations about such things as appearance, posture, and details about his life and crime. He categorized the cards according to the lengths and widths of the criminals' heads and lengths of two of their fingers. For more than two months, he worked feverishly on the project but failed to find a single person who had been arrested twice. Into the third month, after measuring around two thousand men, one afternoon, he came across a man who gave the generic name Dupont but whom Bertillon believed he recognized as someone else. He looked through his cards under the heading "Medium Heads," and pulled one with the name Martin, who had been arrested two months earlier. The measurements and photographs, as well as the physical description, were a match. Bertillon confronted the man, who denied that his name was Martin, but after Bertillon persisted and showed him the evidence, he capitulated. Bertillon was able to confirm for his father that the system worked just before the old man died, and the news spread throughout the local law enforcement community that they had a way of identifying recidivists.

Thus, Bertillon introduced a scientific method into criminal investigations. It was much better than simply running down random clues or trusting in an investigator's memory of faces or voices, and that first year, quite a few criminals were identified. In addition, at times Bertillon was able to give names to deceased John Does whose cards were in his files. He called his innovative technique anthropometry, but it became known as *bertillonage* and quickly became *the* method for identification around Europe. When Bertillon identified an anarchist as a murderer, he earned the Legion of Honor. These were heady days for him and he acquired a great deal of prestige and power.

In 1887, Major McClaughry, the warden at the Illinois State Prison in Joliet, introduced Bertillon's system into America by translating his book into English and using the method in his prison. The practice quickly spread to other American prisons. At the same time, phrenology was becoming popular in the States as well, with a new wave of supporters. One of the key publishing houses for books and pamphlets on the subject was founded by two brothers, Lorenzo and Orson Fowler in New York. They supported the devices invented to assist in taking the skull measurements, including the automatic electric phrenometer.

While all of this was going on, entomology entered another case. Dr. Paul Brouardel, a physician in Paris, autopsied the mummified corpse of an infant, requesting the assistance of Pierre Mégnin, who was a professor at the Natural History Museum. Mégnin examined the larvae on the corpse, identifying mites, moths, and other biological specimens, concluding that the infant had been dumped where it was found between five and seven months earlier. (Mégnin would continue to collect more data from such cases until he published a definitive book in 1894, *La Faune des cadavers: Application de l'entomologie a la medicine Legale*, showing how entomological data could assist in forensic investigations.)

THE SEXUAL CRIMINAL

Psychiatry, too, made progress, and one alienist had been at work on a way to categorize criminal behavior and sexual deviance. The early psychiatrists believed that lawyers needed assistance from psychiatry to understand the criminal mind for greater fairness in legal proceedings, but they realized that their discipline offered little by way of standard ideas and treatment. The legal system already viewed scientific experts with a certain degree of cynicism and the physical sciences were much more precise. One of the alienists who undertook to examine mental illness toward the goal of categorizing it was Richard Freiherr von Krafft-Ebing, a German neurologist who was director of the Feldhof Asylum in Graz, Austria, and a professor of psychiatry at the University of Graz. He served for many years as a psychiatric consultant to the courts, and he eventually operated his own clinic. Thus, he had a good base from which to collect case histories.

He agreed with his colleagues throughout Europe and America that without a standard diagnostic system, psychiatry could not consider itself equivalent to the field of medicine, so in 1880, he published three volumes collectively titled *A Textbook of Insanity,* in which he outlined an elaborate system for categorizing mental diseases. Since insanity had already been treated as a legal concept in England, this medical use of it would cloud the waters, because it would become apparent in certain proceedings that some people who suffered from psychosis might still appreciate that what they were doing was wrong. Thus, they might be medically insane without being legally insane. That confusion still exists to this day.

The next significant text by Krafft-Ebing, published in 1886, was *Psychopathia Sexualis with Especial Reference to the Antipathic Sexual Instinct: A Medico-Forensic Study.* His approach was to identify a foundational problem, the development of degeneracy, and study it via its peculiar manifestations in sexual deviance. He set up a theoretical framework through which to identify and interpret the various behaviors, relying on such items as heredity, corrupting influences on the nervous system, the evolution of a motive, and a qualitative set of details about the personality. In six chapters, he laid out forty-five cases that focused largely on violent criminals or extraordinarily perverse practices. These were the harmful consequences of a degenerate lifestyle, which itself was often influenced by specific types of temptations. Such persons were not well-equipped to resist; they might be timid, lacking in education, or of limited intelligence. Nevertheless, they were responsible for exposing themselves to situations in which their weakness would undermine their efforts to be good.

Krafft-Ebing found in his extensive study a close link between erotic lust and the impulse to murder. By selecting cases to correspond to a simplified framework that discounted multiple motives, he offered psychiatry a "vocabulary of perversion" and a seemingly viable standard of interpretation. It did not necessarily correspond precisely to reality, since it imposed artificial structure on human experience, but it became a key framework for later systems of categorization.

In the preface to the first German edition, Krafft-Ebing states, "Few people ever fully appreciate the power that sexuality exercises over feeling, thought, and conduct, both in the individual and society." To that point, he said, the "empirical psychology and metaphysics of the sexual side of human existence rest upon a foundation that is scientifically almost puerile." His entire purpose for his undertaking was not to offer a psychology of sexuality but to describe the pathological manifestations of the sexual lives of human beings for legal purposes. The subject demanded to be studied scientifically, because the medicolegal expert had to pass judgment, and it would be better for justice that these judgments be solidly based on professional standards. In order to preserve the study for professional reading, he used a title that would be understood only by the learned and wrote the more graphic passages in Latin. Krafft-Ebing did not want the layperson exposed to these lurid descriptions of cannibalism, necrophilia, coprophilia, and lust-murder.

Among his cases were the following: a man whose erotic pleasure lay in gathering pubic hair from the beds of prostitutes, an adolescent who cut pretty girls for sexual pleasure, a prostitute who walked on her hands and got monkeys to undress her, a club for women-haters, and all manner of bondage practices and fantasies. Most notorious were the cases of serial murder that involved cannibalizing the flesh of victims or sleeping with corpses, but more cases were akin to the twenty-year-old boy arrested for stabbing or slapping women in the genitals to ease his lust for female buttocks.

This book was a huge commercial success, despite Krafft-Ebing's attempt to limit it to a scientific audience, and he became a celebrity to the social elite and their literary circles. He also influenced how novelists developed their criminals. *Psychopathia Sexualis* sold thousands of copies to people who had only a prurient interest in the contents, and many strained to figure out the secretive Latin passages. It was translated into English in 1892 and has gone into many printings, as well as new editions that added cases and refined the concepts. In its current and final version, it presents 238 cases. Krafft-Ebing is credited with clarifying terms such as necrophilia, masochism, and fetishism. After writing his book, he was able to learn how it had assisted others in making legislation and jurisprudence more rational by helping to diminish superstitious ideas about sexual disorders. Krafft-Ebing's work also helped to draw a line between normal and abnormal, at least within the context of his restrictive Catholic upbringing. He might have shuddered at the idea that his work foreshadowed an approaching age of sexuality.

FIVE

ANATOMIES OF MOTIVE

INFLUENTIAL SLEUTHS

In 1876, a young man named Arthur Conan Doyle entered the University of Edinburgh in Scotland, noted for its progressive ideas about medicine, to pursue his medical studies. During his second year he clerked at the Royal Infirmary for an impressive figure, Dr. Joseph Bell, a lean, fortyish professor with an aquiline nose and piercing eyes who was also the personal surgeon to Queen Victoria when she came to Scotland. An amateur poet as well, Bell instigated systematic lectures for nurses and edited the *Edinburgh Medical Journal* from 1873 until 1896. He was a firm proponent of science, applying the methods of observation and clear thinking wherever he saw the need.

As Bell walked energetically around the lecture room, he impressed upon his students the importance of performing a close and critical study before diagnosing any situation. He apparently possessed the ability to make accurate judgments in his outpatient clinic from mere observation and he urged his pupils to develop this skill. To demonstrate, he would select someone he did not know and describe details about that person's occupation and recent activities from deductions derived from subtle clues, such as the type of clothing worn, the posture or style of walking, or the presence of calluses or stains on certain fingers. Convinced that the nuances mattered, Bell relied on his eyes, ears, nose, and hands to diagnose diseases. Nothing was more useful to medical work, he asserted, than finely honed sensory observation, guided toward a specific purpose. Later, he would use this skill in a forensic context, but not while Conan Doyle was at the school, unless he consulted behind the scenes in the Chantrelle case, as some scholars suggest.

Eugene Marie Chantrelle was a French immigrant in Edinburgh who had married a sixteen-year-old girl, apparently to dominate her and feed his own ego by making her feel worthless. He liked to brag that he could poison her and in 1877, he insured her life for a considerable sum, to be paid in the case of accidental death. It seemed that she then died from exposure to coal gas leakage, but a noted toxicologist, Henry Littlejohn, thought her symptoms were consistent with poisoning and after his examination he found a high quantity of opium in her vomit. As well, the gas leak was the result of deliberate sabotage, so Chantrelle was arrested. With Littlejohn's assistance, he was then convicted of murder. In light of earlier scandals, this proved to be a positive case for toxicologists.

Since Joseph Bell was a close associate of Littlejohn's, some scholars surmise that he helped with the analysis. He insisted in the classroom that a physician's duty was to consider minutia that most people would view as trifling and to understand the wealth of information it provides. Apparently Conan Doyle absorbed this lesson well.

He'd gone to Edinburgh from Southsea, England, when he was eighteen and remained there (with a brief excursion as a ship's surgeon) until he finished his studies in 1881. Sometime after he left Edinburgh, as he started his own medical practice back in Southsea, he pondered writing a novel. He'd already published a number of short stories, but he sought a more marketable and enduring enterprise, especially as he'd recently gotten married and hoped to raise a family. He thought of writing about a detective, and was not the first author to have done so, although few had paid much attention to what was considered by that time the science of investigation. Conan Doyle certainly knew about anthropometry, the analysis of blood, the early attempts to examine hair and trace elements, ideas about fingerprints, and the improvements in both photography and microscopy. He adopted Bell's approach to give his detective the keen powers of observation and deduction, but he added his own interest in the latest cutting-edge technologies. In fact, the fictional Holmes sometimes describes inventions of his own that predated the actual scientific discoveries.

In addition, Conan Doyle knew about the French detective Vidocq, and had long been influenced by the writings of Edgar Allan Poe, who'd created his own genius detective, M. Auguste Dupin. Poe had set up the "tales of ratiocination" to involve readers in thinking through the mystery at hand along with the investigator. Both were required to use logic and deduction to make the right sort of sense of the clues. As well, Poe employed a sidekick as the detective's chronicler, who provided readers with appropriate background information. Dupin worked alone rather than under the auspices of the police, and was often at odds with them on cases. The first such tale, "The Murders in the Rue Morgue," had been published over four decades earlier in 1841, and it introduced the element of pitting the detective against a seemingly impossible puzzle, which only the combination of keen observation, an open attitude toward the seemingly improbable, and a sense of deductive intuition can finally solve.

Also in Russia in 1866, Fyodor Dostoevsky had created Inspector Porfiry in *Crime and Punishment*, who must solve the murder of an elderly pawnbroker and her stepsister, apparently committed by someone who had not even bothered to rob them. Given all the people who had business with the victims, the task of narrowing leads is prodigious, and eventually it falls to psychology rather than physical evidence analysis to solve it. Porfiry discovers that a student named Rodión Raskólnikov has published an article on the idea that extraordinary men can commit crimes without moral accountability. Porfiry pressures Raskólnikov, and his inability to actually be the remorseless extraordinary man finally triggers a confession.

Thus, Conan Doyle set out to follow in a tradition already pioneered by others, adding some ideas to his stories that he'd mentioned in his notes. There were as yet no texts on criminology or criminalistics, although there had been some in fields like entomology and toxicology, and aside from the authors who had utilized Vidocq's methods in their fiction, there were few descriptions in print about specific forensic procedures. In 1886, with the keen-eyed observer in mind, Conan Doyle completed a story about what he called a "consulting detective" named Sherlock Holmes, who resided at the fictional 221B Baker Street in London. Holmes's companion, Dr. John Watson, duly chronicled the master sleuth's cases. The first of what would become a series of tales was *A Study in Scarlet.* Initially titled *A Tangled Skein* and featuring "Sherrinford Holmes," this tale was rejected numerous times before Ward, Lock & Co. published it in 1887 as part of *Breeton's Christmas Annual*, and now it starred Sherlock Holmes. It attracted little attention but when it was published on its own in 1888, it received greater exposure.

The story opens with a murder that stumps detectives at Scotland Yard, so they call Holmes in as an independent consultant. He realizes that the official hypothesis about the incident is clearly wrong and initially interprets it as having a political motivation. This first tale was cognizant of the developments in physical anthropology, as Holmes emphasizes from items at the scene significant revelations about a perpetrator's physical stature and appearance. In fact, as a starting point, he notes a fingerprint. During this era, many experts believed that fingerprints might identify people of certain races. To continue, five letters have been scrawled on the wall by a blood-covered index finger. From the position he deduces the offender's height, and from other details of the lettering (shape, breadth of the lines), he deduces his stature and likely nationality. Given how much blood it required to write the message, Holmes supposes the man has a "florid" complexion. He also has small feet, smokes a specific type of cigar, and has fingernails of a certain length. As he gains more information, Holmes reinterprets these clues to decide that the crime is actually personal in nature.

His next case, published in 1890 as *The Sign of Four*, involved reading "footmarks," which showed everything from the height and weight of the perpetrator to his psychological state. Conan Doyle poses a person's physical make-up as an encoded text that only the most astute investigator knows how to read. In fact, Henry Faulds's article describing fingerprints found on Japanese pottery as a racial identification had recently been published in the science weekly, *Nature*, so it's likely Conan Doyle was aware of it.

Fans of Sherlock Holmes wanted more, so he penned more, and the series would successfully run, gathering many more readers, until 1927. Among the reasons why the Holmes character endured was the technical expertise in the tales and the surprises that emerged from careful analysis, as well as the explanations Holmes offered of the thinking process. Once readers could see how he arrived at his deductions, they believed anyone with a capable eye could do what he did, so he became a role model. As the various scientific disciplines merged with crime investigation during this time, Holmes became the prototype of the brilliant freelance detective.

In the more realistic world of crime, the science of identification continued to develop, but it would have to compete for media attention with a series of incidents that placed an international spotlight on the weaknesses of British law enforcement.

URGENT INVESTIGATION

In the same year that Holmes emerged into public awareness, Eduard Charles Godon, a dentist who had established the Paris Odontological Society in 1882, recommended using dental records to link unidentified remains with missing persons. In 1888, Bertillon took over the new department of identity in Paris, introducing the profile angle into mug shots (*photographie métrique*), which offered the jaw, ear, and more of the face for identification. He also required more precision for crime scene photos, insisting on getting images before the scene was disturbed and photographing them in a way that one could reconstruct how evidence was placed at the scene. He had mats printed with metric frames to mount on the photos, and for some he included both front and side views of specific objects. Yet even as he managed all of this, Scotland Yard was about to face its greatest challenge.

Its criminal investigation department had been created a decade earlier, and Superintendent Charles Vincent soon released rules for dealing with murder cases, which stated that "the body must not be moved, nor anything about it or in the room or place interfered with, and the public must be excluded." But for the spate of murders they had to process in the Whitechapel slum toward the end of 1888, they were hardly prepared.

On Friday, August 31, just after 1:00 A.M., Mary Ann "Polly" Nichols offered herself for lodging money, but her potential customer slit her throat instead, using two quick strokes, slashing her abdomen, and dumping her in the street. Only three weeks before this incident, a prostitute named Martha Tabram was stabbed thirty-nine times, but the detectives did not think the two murders were related.

The next "official" victim was Annie Chapman, discovered killed in the morning hours of September 8. Her dress covered her head and her stomach was ripped open, with her intestines exposed. Her throat had been cut with what appeared to have been a sharp and narrow surgical knife, but, oddly, several coins and an envelope had been arranged around her. An autopsy revealed that her bladder, half of the vagina, and the uterus had been removed. Since these items were not at the scene, investigators believed that the killer had taken them.

There were no collective clues from these victims that pointed to a man the women might have tangled with, and thus no clear leads. Prostitution was a high-risk business, but this kind of slaughter was unusual. At least two of the women had been attacked quickly and out in the open, but no witnesses came forward to offer descriptions. But about three weeks later, a note arrived to the Central News Agency on September 27 that began, "Dear Boss," and was signed, "Yours Truly, Jack the Ripper." Its author claimed that he "loved" his work and would continue to kill. On September 30, there were two victims on the same night, both murdered outside.

Despite the "rules" for investigation, it seemed clear that there was little regard to the careful handling of bodies and their effects, or of even preserving the crime scenes. Indeed, some evidence was poorly handled and untrained mortuary workers sometimes removed the victims' clothing. They failed to label or preserve anything for potential use in an arrest and trial. Even Sir Charles Warren, in charge of the Metropolitan police, rubbed away a message at the crime scene, apparently written after one of the murders, for fear it might incite an uprising. He could at least have taken a photograph first.

Two weeks after the "double event," the leader of the Whitechapel Vigilante Committee received a letter and a box "from Hell." Inside the box was half of a kidney preserved in wine. The note's author indicated that he'd fried and eaten the other half, which was "very nise." It was believed that the killer sent this note, signing off with, "Catch me if you can."

But the police faced another problem: They were receiving numerous correspondences that claimed to be from the killer but were clearly faked. It was difficult for them to pull an authentic letter from the lot to assist with leads. They had never before been faced with such an investigation.

On October 3, as the New Scotland Yard was being constructed to house a growing metropolitan police squad, someone deposited the headless, limbless corpse of a woman inside a vault in the unfinished foundation. A police spokesman told journalists that it was not considered part of the murder spree.

Twenty-four-year-old Mary Kelly was the final official victim. This time the Ripper had accompanied her into a room and after slitting her throat, mutilated her remains. When police arrived the following morning, they found the room spattered with blood, brains, and pieces of internal organs. Dr. Thomas Bond, a surgeon who assisted in the autopsy, stated that the murders had escalated in brutality and were sexual in nature, with an intense element of rage against women or prostitutes. Bond listed certain traits to be expected from this man, including that he was physically strong, calm under pressure, and daring. In ordinary life, he was likely quiet and apparently inoffensive, middle-aged, and neatly attired. It seemed likely that, to hide blood on his clothing, he would wear a cloak. To get people to provide more clues, Bond suggested that the police offer a reward.

On July 17, 1889, a constable found a woman who had just been murdered; her body was still warm to the touch. She was identified as "Clay Pipe Alice" McKenzie, a heavy drinker and possible prostitute. Bond offered a second opinion after the original autopsy by Dr. Bagster Phillips, concentrating on the stab wounds to the neck.

They were not the Ripper's typical slashing method, but there had been incisions in the abdomen. While Bond insisted that this was a Ripper victim, Phillips believed just the opposite; in fact he did not think that the earlier five victims had all been the work of one person, and he claimed that he'd arrived at this conclusion "on professional grounds." He had performed five of the seven autopsies, and without including the circumstantial evidence in his calculation, he made his judgment from the bodily wounds alone.

While the detectives from Scotland Yard were roundly criticized for their inability to identify and arrest the killer, they did keep some suspects under surveillance. They also surveyed their approach, resolving to tighten it up, rely more on science, and become more rigorous in their methods.

BACK TO FINGERS

Many problems arose among those police agencies relying on *bertillonage*, notably that not everyone taking the measurements was careful or patient with its time-consuming demands. Bertillon was a perfectionist and while he took pains to teach others how to apply the techniques, he could not be everywhere at once. Even as he attempted to retain control, the method of analyzing fingerprints gained momentum. Some sources claim that Bertillon opposed it, but others indicate that he was pleased to add fingerprints to his own card system, although he did not see how they could be used to categorize the cards the way his system could. In any event, there were too many people in separate places taking note of the value of fingerprints for the method to remain dormant for long.

Francis Galton had studied *bertillonage* up close, being instructed by Bertillon himself, but when he decided the system was too complicated, he went to see Herschel in England to learn more about fingerprints. Like a true scientist who cared about advancing knowledge, William Herschel offered everything he had, hopeful that Galton could take it further than he'd managed to do. Galton set about studying the method in a painstaking manner for a period of three years, as well as trying to figure out a way to systematize it. He found Czech physician Johannes Evangelista Purkinje's technique from earlier in the century too complicated and believed there must be an easier way. He published a paper in *Nature* in 1891, which—not surprisingly—drew a venomous reaction from Henry Faulds, claiming again that *he* was the inventor of the fingerprint technique. Galton ignored this rivalry and continued with his work.

The following year, he published the first book about fingerprints and their forensic utility, simply titled *Finger Prints*. He proposed that fingerprints bore three primary features and from them he could devise sixty thousand classes. But he was stuck at completing a practical system and he knew there were flaws. He needed help, and like Herschel before him, was eager for a collaborator.

Across the ocean in La Plata, Argentina, an enterprising investigator had already made this method work. The chief of police, Guillermo Nuñez, had learned about the idea of matching fingerprints to offenders and he charged Juan Vucetich, head of the statistical bureau, with implementing it. Vucetich soon had the opportunity.

In 1892, in the small town of Nocochea, Francisca Rojas claimed to have found both of her children, ages four and six, brutally bludgeoned to death. She accused a man she knew named Velasquez, who wanted to marry her and who had threatened her when she said no. He protested his innocence, so the police chief required Velasquez to spend the night next to one of the corpses, with the hope it would rattle him sufficiently to start talking. But there was no confession, not even when torture was applied.

But rumor had it that Rojas had a young lover who had declined to marry her because of her children. Now *she* had a motive, along with a reason to frame someone else, so the chief turned to a rather unorthodox technique: He tried to frighten her into a confession by making ghostly sounds outside her home. This, too, advanced the case no further.

By then Vucetich had formulated his own system from fingertip ridge patterns for identification. He had told Chief Eduardo Alvarez, the official in charge of this investigation, about it so when the chief found a brownish smudge in Roja's home that was clearly a bloody fingerprint, he compared the pattern with both suspects and identified the mother as the person who had placed her thumb in blood. That indicated that she had murdered the children. When she was told about this evidence, she broke down and confessed that she had killed them both with a rock. Velasquez was freed and Rojas convicted.

Vucetich went on to identify more people with this method. He wrote two books describing "dactyloscopy" and defending the system as superior to all others. He nevertheless traveled to England to meet Bertillon, who remained aloof in his inimitable and annoying manner, refusing to meet with him. By then, fingerprinting dominated others in Argentina as the method for criminal identification. Still, it lacked a system, and it remained for Edward R. Henry, a magistrate in India, to finish the job.

Henry had tried a simplified method of anthropometry, but finding it cumbersome, had then looked to fingerprints as the source for identification. Independent of the developments in Argentina but influenced by Galton's work, Henry had started a fingerprint classification system as inspector-general of police in Nepal, India. He traveled to England to show someone, and Galton proved eager to offer him materials and support. Henry saw the problems that had stymied Galton but did not immediately see a solution.

His inspiration occurred while looking at his own fingerprints during a train ride in 1896, and he soon devised a loop and delta system. He separated fingerprints into two basic groups: value patterns (whorls) and nonvalue patterns (arches and loops), and described how the lines could be counted for individualizing them within these categories. With assigned values for different fingers, he formed codes for each set of prints. Thus, the prints were filed via their numeric codes and could easily be retrieved for comparison. This process decreased the time heretofore involved to look them up. However, the system required all ten prints for an identification and crime scenes often turned up only a few—even just one. But this issue had not yet become a major problem and professionals were starting to encourage others to use this system. Nevertheless, reliance on fingerprints as proof of identification needed a defining case, which was still years away.

In the meantime in France, another forensic expert had been busy making a different kind of mark.

CRIMINALISTICS

In 1885 in Lyon, France, an elderly man was found dead on the bed in his locked bedroom, with a wound on his skull and his hand firmly grasping a pistol. There was no evidence that someone had forcibly entered the room, so two physicians called to the scene stated that the manner of death was suicide. It seemed obvious. However, when professor of medicine from the University of Lyon Dr. Jean Alexandre Eugène Lacassagne examined the scene, he was unwilling to just accept the obvious. In fact, he noted something odd: The bed linens covered the dead man's arms, a difficult feat for someone who has just put a bullet through his brain. Upon closer inspection, Lacassagne noticed other indications of staging: the eyes were closed and the typical gunpowder burns on the skin from a close-range shot were lacking. Before making any statements, he decided to do some research.

Lacassagne believed that it was imperative to use observation and critical examination to arrive at the big picture before he made a final judgment. He knew from his experience working on corpses how the typical suicide-by-gunshot would appear, and he was convinced from the appearance of the wound that the weapon that had killed this man had been too far away for him to have managed it. Yet Lacassagne was not certain about the relationship between violent deaths and open or closed eyes, so he questioned a number of nurses. They said that only in natural deaths had they seen the eyes closed; in contrast, sudden or violent deaths generally resulted in the eyes remaining open, even staring. Assured that he was on the right track, Lacassagne turned his attention to the revolver in the dead man's hand. Such a tight grip would be difficult to achieve by placing the weapon there after death, but he considered several possible scenarios and then collected data from an experiment.

He requested that other medical personnel inform him immediately whenever someone died in the area, and he would then go to the scene to attempt to place an object into the decedent's hand. He learned that a dead hand, immediately postmortem, could indeed be made to grasp something like a gun, though not tightly. However, once the early stages of rigor mortis set in, a gun placed in an initially loose grip could become more tightly clasped. Relying on the evidence, his experiments, and his own reasoning, Lacassagne decided that the elderly man had been murdered. The police turned their attention toward the victim's son as an obvious suspect. It seemed that while he'd hoped to dispatch the man, the slight degree of fondness he felt had compelled him to close his father's eyes and cover him. The son was tried and convicted on Lacassagne's findings, yet had the police relied only on the doctors who had performed a cursory examination, this killer would have gotten away with his crime.

In his day, due to his careful work, prestigious university position, and authoritative bearing, Lacassagne became a celebrity investigator. Born in 1843, he had attended military school and qualified as a surgeon, becoming an army physician in North Africa. There he'd developed an interest in pathology and the identification of the dead, and was especially adept at the interpretation of wounds and casualties from violent incidents. In 1878, when he was thirty-five, Lacassagne published *Précis de Médicine*, which helped him to obtain the teaching position at the University's institute of medicine, which was established two years later.

He was an obsessive learner who read documents from a variety of fields, and because his knowledge proved comprehensive he won the respect of colleagues in other disciplines as well. Yet he remained ever cognizant of the limitations of medicine, and among his cautionary practices was to exercise doubt about even the most seemingly obvious situations, because one might otherwise miss the tiny clues that could produce a more accurate opinion. By the time he retired in 1921 and donated his books, he had an impressive collection of over twelve thousand volumes.

Lacassagne's interest in criminology had a long history. A once-avid student of the works of Italian criminal anthropologist Cesare Lombroso, who believed that criminal types could be recognized via bodily "maps," Lacassagne redirected the discipline by noting sociological influences on criminal behavior. He formed a group of professionals around his own ideas, becoming famous for the comment "Societies have the criminals they deserve." While he believed that disease and addiction, passed on to successive generations, could cause mental and physical degeneracy, poverty, social marginalization, and other factors were also involved. In fact, in a speech given in 1881, he stated that the fight against criminality was one of the physician's social responsibilities. "At the present time, it will be the physicians, once again, who will show judges that some criminals are incorrigible [and] some are organically bad, defective individuals . . ." Lacassagne came to realize that, contrary to the ideas of the physical anthropologists, criminals appeared physically normal but were vulnerable for various reasons to corrupting social influences. "The criminal is a microbe," he said, "that proliferates only in a certain environment." He then launched a journal, *Archives de l'Anthropologie Criminelle,* to discuss social initiatives to ease crime.

He also studied tattoos. Lombroso had made the observation that a certain type of person, "locked in combat with society," resorted to decorating his body with a tattoo, and different designs were indicative of different temperaments in that type. The more obscene and the more percentage of the body involved, the less sensitivity to pain and the greater the tendency to be savage. Lacassagne was more careful as he assembled a large collection of thousands of examples, categorizing them and seeing in them the desire to express ideas in symbolic form. They spoke to something about culture rather than necessarily to savagery or criminality.

Lacassagne's most memorable contributions derived from his work in pathology, notably his observation of the stages of death. Many nineteenth-century pathologists believed that the time of death could easily and accurately be determined by measuring body temperature and stages of rigor mortis and lividity, but the more homicide cases they saw, the less certain they were. It took observant physicians such as Lacassagne to recognize and admit that the interpretation of decomposition and other postmortem indicators was not sufficiently reliable to be deemed a science. There were too many environmental and individual variables. Nevertheless, he spent considerable time making calculations from the dead to better understand the postmortem interval.

Lacassagne also made a contribution to several other areas of investigation, including the field of ballistics. After removing a bullet from a victim during an autopsy in 1889, he noticed longitudinal grooves on its surface and counted them; there were seven. Then he examined the barrels of several pistols that belonged to the various suspects and identified the one he believed had been used to commit the murder—the only one with seven grooves. Its owner was convicted. While this kind of analysis was primitive and could easily have been wrong, the early scientists—mostly physicians, pathologists, and chemists—were at least moving in the right direction.

One of the first cases in which hair was carefully analyzed also involved Lacassagne. In August 1889, the decomposed nude body of a bearded, dark-haired man was found near Le Tour de Millery, France, ten miles from Lyon. It had been wrapped in oilcloth and placed head-first inside a canvas bag. Dr. Paul Bernard, who had once studied with Lacassagne, examined the corpse with some difficulty, since it was in an advanced state of decomposition and had a terrible odor, but he estimated the victim's age to have been around thirty-five. He could not be certain, but he thought the man had died by strangulation, since the larynx showed two breaks. Around the same time, a wooden trunk, broken into pieces and smelling distinctly of decomposition, turned up not far away, with a shipping date from Paris that was difficult to read. Investigators speculated over whether it might be associated with the deceased, and a key found near the area where the corpse was dumped fit the lock, so that clinched the connection. They started to work on learning when it had shipped.

The corpse was not identified as a local resident, so the press picked up the story, and it reached the ears of Assistant Superintendent Marie-François Goron at the *Sûreté* in Paris. He looked through his reports on missing persons and came across the name of Toussaint-Augsent Gouffé, a known womanizer whose brother-in-law had reported that he had not been seen since July 27. Gouffé was a forty-nine-year-old court bailiff with an office in the neighborhood of Montmartre, and during the investigation Goron had learned that a strange man had entered

Gouffé's office on July 26 and left in a hurry, without stating his business. Since Gouffé's financial affairs were in order and there appeared to be no reason that he might flee town or kill himself, Goron paid more attention to the case.

He sent Gouffé's brother-in-law to Lyon to look at the remains found there, which were stored on an odiferous barge converted into a morgue. The man looked at the black-haired, rotting corpse for as long as he could bear it and stated that because Gouffé had chestnut brown hair, it was not him. The unidentified decedent was finally buried, while Gouffé remained among the missing.

However, Goron thought there were too many circumstances in common for the cases to be unrelated, so he persisted. Then a Lyon cab driver claimed to have picked up a heavy trunk from the railroad station on July 6, along with three men who had asked him to take them to the vicinity of Millery. They'd left the trunk there, returned to Lyon, and were arrested for robbery. Thus, since this all occurred before Gouffé had disappeared, the corpse that had clearly been in the trunk could not be the missing Parisian. But Goron was not convinced that this ended the case, because in his experience, witnesses could be mistaken. He learned that Gouffé had been seen with a man named Michel Eyraud, a known pimp and scam artist, and Eyraud's mistress, Gabrielle Bompard. Both had also vanished from Paris on July 27. Then the police managed to track the trunk to a shipping agent in Paris whose records indicated that it had been sent to Lyon on July 27. Goron was now certain that the corpse was their missing citizen, especially after the cab driver admitted he had fabricated his tale about the trunk.

Goron asked Dr. Bernard for a sample of the dead man's hair, kept in a test tube, and immersed it in distilled water. The "black" hair, now free of grime and crusted blood, turned out to be chestnut brown, the color of Gouffé's hair, so Goron petitioned for an exhumation and invited Lacassagne to make an opinion. (Some sources indicate that Lacassagne is the one who washed the hair.)

Knowing that a bungled autopsy cannot be revised, Lacassagne performed a reautopsy as best he could on November 12, 1889, removing the putrid flesh to examine the bones. He noted with dismay that Bernard, his own former student, had smashed the skull and sliced so ineptly through the flesh that much of it was damaged. It took Lacassagne eleven days to make his way through the putrid remains. His colleague, Dr. Etienne Rollet, had recently published a method for determining the size of a body from the bones, so Lacassagne relied on these calculations to state that the man had once been about five-foot-eight, which matched information on Gouffé's military record. Under a microscope, Lacassagne compared hair removed from Gouffé's hairbrush with selected hairs from the corpse. Their thickness corresponded exactly, and another type of analysis showed that no hair dye had been used. In addition, the corpse had a deformity in the bone of the right knee from an accumulation of fluid, which corresponded to a limp that the missing Gouffé was known to have (and his doctor confirmed treatment for water on the knee). Wear on the dentine of the teeth and an accumulation of tartar revealed a man closer to fifty than thirty-five, and Gouffé had been forty-nine. Lacassagne also reexamined the bones of the throat and found the two breaks, which confirmed some type of strangulation as the cause of death, and he believed it had been done manually. On November 21, Lacassagne said to Goron and his associates, "Messieurs, I herewith present you with Monsieur Gouffé."

Once Gouffé's identity was established, the police set out to find his killer, and it proved to be an ingenious bit of work. Goron had a replica of the steamer trunk built and placed in the Paris morgue for the public to see. Twenty-five thousand people filed past. When no one offered useful information, a photograph was published in newspapers abroad. A Frenchman in London who saw it recalled meeting a father and daughter who had purchased one like it. This information was publicized as well, and Goron subsequently received a twenty-page letter from Michel Eyraud, then in New York. He accused his mistress, Gabrielle Bompard, of the murder, and she arrived in Goron's office to implicate them both in the grifting scheme. They had killed the man, she admitted, attempting to hang him but finally strangling him with their hands (as Lacassagne indicated). Their trial was brief and both were convicted. Eyraud was executed and Bompard given a prison sentence.

During Eyraud's execution, peddlers sold tiny replicas of the trunk with lead corpses inside, bearing the inscription, "L'Affaire Gouffé." This case was widely publicized, turning forensic pathology into a public sensation. The careful unraveling of the mystery and the collective persistence of brilliant minds gave the struggling discipline of forensic science a real boost.

After his involvement in this incident, Lacassagne's fame spread internationally, so people listened when he introduced new ideas into other forensic arenas. He analyzed bloodstains at crime scenes and studied the criminal psyche. He was also invited to analyze criminal suspects and offer his opinion. In 1897, a tramp from southwestern France named Joseph Vacher was accused of crimes against fourteen people, including eleven murders. The police also believed he had raped as many as forty children. Vacher had been arrested after a seventeen-year-old shepherd was found strangled and stabbed, with his belly ripped open, and Vacher, twenty-nine, offered a written confession in which he claimed to suffer from an irresistible impulse that drove him to commit murder. Having been bitten by a rabid dog when he was a child, he insisted that his blood had been poisoned. As his victims died, he admitted, he drank blood from their necks.

A team of doctors, including Lacassagne, examined the defendant for five months, learning from relatives and associates about his "confused talk," spells of delirium, persecution mania, and violent history. Indeed, three years earlier he had been treated in an asylum after he killed a woman and had sex with her corpse, and it was also known that he'd removed the genitalia from several children. Lacassagne nevertheless decided that Vacher was faking a mental illness. Because the assailant's memory was clear about the crimes and because he had run off after committing them, he had demonstrated sufficient awareness to be judged sane and thus responsible for what he had done. In 1898 in court, Lacassagne demonstrated how he believed the defendant had carried out the crimes and Vacher reportedly said, "He's very good." Indeed, Lacassagne's reputation and commanding stance helped to convict Vacher, who was executed two months later by guillotine.

In addition to making analyses, Lacassagne also instigated the earliest criminal autobiographies. He encouraged a number of prisoners to write about themselves, and each week he checked their notebooks, correcting and guiding these men and women toward some revealed insight. He found that their family histories were full of violence, tension, and disease. Interestingly, these prisoners seemed eager to contribute to the new science of the criminal type and some even critiqued the theories. Like criminals today, they seemed aware of what was being said about them.

While Lacassagne made his careful studies in France, Austrian lawyer Hans Gross, a longtime friend of Richard von Krafft-Ebing's, became versed in a variety of sciences from physics to psychology, and would make a contribution to forensic science that would prove to be among the most influential in the field thus far. His enthusiasm for science, deriving in part from his fascination with the marks on a bullet removed from his grandfather, was contagious and he pursued a way to integrate everything he knew into a comprehensive package that would benefit others.

Gross coined the word *criminalistics* and founded the Criminalistic Institute at the University of Graz, offering it for the collaboration of diverse specialists involved in police science. He collected a library (probably including British physician and social reformer Havelock Ellis's new book, *The Criminal*, about criminal anthropology). In 1891, Gross published *Handbuch für Untersuchungsrichter* and two years later *System der Kriminalistik* (translated in 1907 into English as *Criminal Investigation*), the first comprehensive description of the practical analysis of physical evidence such as blood, trace evidence, bullets, and fingerprints in solving crime. Among the topics were the criminal use of deception and disguise, how to reveal a con, the rules of investigation, the requirements for good detectives, the art of microscopy, and the need for recognizing secret hiding places when investigating a crime. He also emphasized the use of minute particles of dust as evidence, because many different occupations left distinct traces.

Gross urged investigators to be calm and to refrain from violating the crime scene. Before anything is touched or lifted, it first must be fully described, sketched, or photographed. The important thing, he stated, was to collect accurate data before making a judgment. Even the smallest items might influence the solution, and thus everything transported for further study must be carefully handled so as not to damage or contaminate it. If there were body parts or fluids, they should be placed into separate containers to keep them from mingling. Even in the mortuary or lab, he added, rules should be established and strictly observed. For the detective who had to devise a reconstruction, he must not only record what clearly occurred but also note what did not happen. A "burglar" broke in, for example, but the watchdog did not bark.

As the nineteenth century drew near a close, thanks in part to careful men like Gross and Lacassagne, some detectives were taking pains to preserve crime scenes, handle evidence, and analyze an offender's modus operandi. Nevertheless, certain incidents challenged even the best investigators. Some criminals were aware of the contributions of science, and the most inventive offenders sought ways to conceal their crimes from the investigative probe. Even before the century turned, there were still a few sensational cases to challenge the latest forensic ideas.

SIX

VOICES OF AUTHORITY

CHALLENGES

Robert Buchanan, a practicing doctor in New York since 1886, was involved with the wealthy and much older Anna Sutherland, a brothel madam, who married him and put him in her will as primary beneficiary. In 1892, after they argued over how Buchanan wanted to run his life and whether he would remain in Anna's will, she mysteriously died and Buchanan was suddenly quite rich, to the tune of $50,000. The physician who looked at the corpse noted the dilated eyes, so he decided on a brain hemorrhage as the cause of death.

But then a local reporter, Ike White, sniffed around and learned from Anna's former business partner about suspicious behavior on Buchanan's part—notably that he had remarried his first wife soon after Anna's death—suggesting that he might have poisoned his inamorata.

One of Buchanan's acquaintances even suggested that he'd learned from the proceedings of a recent trial how to administer a morphine overdose. Yet, aware of how morphine constricted the pupils of the eyes, Buchanan had been quite clever. Hoping to prevent doctors from seeing the signs of a morphine overdose, he'd used atropine to dilate Sutherland's pupils. White knew about such treatments, so to confirm his suspicion he interviewed a nurse, who recalled seeing Buchanan put something into his former wife's eyes when she was ill. White printed his speculations, and under public pressure, the coroner exhumed the body. The second autopsy confirmed that the woman had died from a morphine overdose. Professor Rudolph Witthaus, a toxicologist, also affirmed that atropine could dilate the pupils, so Buchanan was arrested.

His trial began on March 23, 1893. Witthaus testified that he had done a color reaction test to confirm the presence of morphine in a lethal amount in the deceased's tissues. The prosecutor picked up from this and demonstrated the atropine technique by killing a cat in court and dropping the chemical into its eyes. It was an unprecedented and dramatic demonstration, and the jury members nodded in approval.

However, such tactics were no longer sufficient. The defense attorney showed with its own expert's exhibit that the color test that Witthaus had used was flawed. Professor Victor Vaughn proved that a cadaveric alkaloid acted just like morphine. That, too, impressed the jury. In addition, in Vaughn's support was the historic discovery two decades earlier by Italian chemistry professor

Francesco Anselmo who had first learned how alkaloids in corpses mimicked certain poisons, such as morphine and delphinine.

In fact, a decade earlier in 1882, this issue had been raised in a British court. Dr. George Lamson stood accused of poisoning Percy John, his brother-in-law, to collect insurance money. The suspect substance was aconite, for which there was still no definitive test. Lamson had visited Percy at school, giving him treats, and two hours later the boy died. Lamson readily surrendered to arrest, believing the pathologist could never detect what he had done, and that was indeed the case. But the *lack* of results from other tests made the doctor suspect aconite and an investigation turned up Lamson's recent purchase of the substance. His defense attorney raised the issue of cadaveric alkaloids, but it appeared to have little influence as Lamson was convicted and sentenced to die. Still, the presence of these substances placed the reliability of the color test in doubt.

Back to Buchanan's trial. While the defense ably proved that they could get the same red color with their tests when morphine was not present, the attorney foolishly decided to put Buchanan himself on the stand. His obsequious manner was so annoying and his story so riddled with significant holes that reasonable doubt soon dissolved. The jury ignored the toxicological arguments altogether and convicted him, based on a combination of circumstantial evidence and his own self-undermining remarks.

Nevertheless, the toxicologists involved realized that

to be more effective in court, they would have to devise a more definitive demonstration for morphine detection—especially when Buchanan's attorneys launched an appeal based on "tainted" scientific evidence. So Professor Witthaus went to work on the problem of the color test's results and tracked its source to impurities in the chemicals used. He was relieved to discover that the test itself was not flawed, and when the appeal was heard, he was ready with his explanation, so the appeal failed, and Buchanan was electrocuted. Witthaus set about updating the color reaction tests to ensure that other toxicologists would not be humiliated in such a manner again.

BLOODY MURDER

On August 4, 1892, the bodies of Andrew Borden, seventy, and Abby Borden, sixty-four, were found in their home in Fall River, Massachusetts, hacked to death. Abby's body lay on the floor of the upstairs guest bedroom, slain with a sharp instrument that had inflicted multiple blows to the back of her head. A thick bunch of artificial black hair, hacked off her, lay nearby. Downstairs, Andrew's corpse, with blows to the face, was half sitting and half lying on the living room sofa. There were blood spatters on the floor, the wall, and the picture over the sofa, but his clothing was strangely undisturbed. Uncharacteristically, his expensive overcoat lay rolled beneath his head. Being a frugal man, it was unlikely that he would have used it as a pillow, especially not on a hot day.

Andrew's youngest daughter, Lizzie, had discovered his body. A thirty-two-year-old spinster, she lived in the small house as well. She immediately called to Bridget Sullivan, the maid, who had gone to her third-floor room to rest and sent her for the doctor.

A neighbor, Mrs. Churchill, arrived and asked Lizzie about Abby. Lizzie said she was out visiting, but mentioned that she'd heard her return and go upstairs. Oddly, Lizzie offhandedly remarked, "I don't know but they killed her, too." Mrs. Churchill was nearly up the stairs when she spotted the body on the guest-room floor.

That someone could have just come in and killed these two elderly people in the middle of the day in a house in the center of town seemed odd, since it was the family practice to keep the doors on the first floor locked. Yet on that day, Lizzie claimed to have been in the barn loft for twenty minutes, leaving the back screen door unlocked—by coincidence the very time that a maniac happened by. No footprints were found around the house on the grass, and no neighbors or workers in the yard next door saw a stranger coming or going in the Borden yard, although a witness reported seeing an agitated young man out in the street, near the Borden fence.

While a search was underway, Sergeant Philip Harrington noticed that Lizzie seemed calm and unemotional, with no stated interest in catching the killer. He had the impression she knew more than she was saying. He saw Dr. Bowen, the physician for whom Lizzie had sent, in the kitchen with scraps of paper on which there was some writing. Bowen said, "It's nothing," but Harrington recalled that one document was addressed to "Emma." Bowen then took the lid off the kitchen stove and tossed the scraps into the fire. Harrington noticed a cylindrical object in the ashes, about a foot long and two inches in diameter. He thought it might have been paper scrolled up.

Another investigator noticed a pail of water in the wash cellar that contained several small bloody towels. He asked Lizzie about this and she said she had explained it all to Dr. Bowen, who said they were menstrual rags, but no one checked to make certain of this. Lizzie said the pail had been there three or four days, although Bridget claimed she had not seen it before that morning. No investigator followed up to clarify this significant contradiction.

Dr. William Dolan, the medical examiner of Bristol County, took possession of all the evidence. At 3:00 P.M., he had the bodies moved into the dining room, where he conducted the autopsies on undertaker boards on the table. He removed the stomachs to help determine time of death, and sent them to Edward S. Wood, professor of chemistry at Harvard Medical School.

Based on the comparative temperatures of the bodies, the varied condition of the blood on each, and an examination of their digestive systems, it was determined that Andrew Borden had died at least one hour after Abby. He was hit in the face, and one eye was cut in half, his nose was severed, and there were eleven distinct cuts. He had been struck from above the head by blows delivered vertically with a sharp implement. Abby had been hit eighteen or nineteen times, probably with a hatchet or ax, with one misdirected blow striking the back of her scalp. The pathologist decapitated the corpses to take the skulls for flesh removal to better see the damage.

Dr. Frank Draper examined the skulls, measuring the location and length of the wounds. The first wound on Andrew had penetrated the nose and upper lips to the tip of the chin, measuring four inches. One of the wounds at the top of the skull gave an accurate measure of the blade, three and a half inches. Besides Abby's many wounds to the back of her skull, she bore a bruise on her shoulder blade that proved to be the shape and size of an ax head.

However, a microscopic examination of the hatchets and axes from the Borden basement produced no blood or hair, except apparently from an animal. They did find a hatchet head without a handle and while the worn blade fit into the skull wounds of Andrew Borden, a new, identical hatchet did not. The head had been forcibly rubbed with ashes, for the crevices in the blade were tightly packed, and investigators surmised that this was an attempt to remove blood and flesh. Yet traces of silver nitrate found behind Abby's ear indicated that a new hatchet had been used.

Lizzie was arrested, and her contradictory answers at an inquest coupled with the grand jury testimony, resulted in jailing her until she could be tried. Implicating Lizzie, blood was found on the sole of one of her shoes and there was a small spot of blood on one of her underskirts, one-sixteenth of an inch in diameter, more extensive on the outside than in. (Apparently there were also bloodlike smears on her skirt that no one tested.) Lizzie said she'd scratched a flea bite. She handed over a dress she said she had worn on the morning of the murder, and it had no blood on it, but some people who had seen her believed it was not the right dress.

The day before the murders, Lizzie had also gone to a drugstore and asked to purchase prussic acid to rid a sealskin cape of bugs. The pharmacist refused to sell it without a prescription. That night, she visited a friend, Alice Russell, and remarked on a strong sense of foreboding that someone was trying to harm her father. She mentioned a man who had argued with her father recently about property. Alice also observed Lizzie burning a "useless, faded" dress in the kitchen stove three days after the murders. Alice's testimony about this incident at a hearing prompted Judge Blaisdell to charge Lizzie with the double homicide.

It was no secret that Lizzie had despised her stepmother. She and her older sister, Emma, had been annoyed a couple of years earlier when Andrew had turned property over to Abby's sister, which the daughters viewed as their birthright. In addition, Andrew was a wealthy man, yet he forced the family to live frugally, and Lizzie was resentful. Also, Bridget Sullivan claimed to have heard a loud laugh coming from the head of the stairs on the fateful morning as she let Mr. Borden in the front door. At that time, Abby Borden was lying on the guestroom floor, so whoever stood on the landing could certainly see her.

At Lizzie's two-week trial, the prosecution, led by Hosea M. Knowlton, made the following points: No witnesses saw anyone enter or exit the Borden home that morning; no "intruder" could count on the coincidence of Bridget being upstairs, Lizzie being outside, and an overnight visiting relative, John Morse, who was out that morning, not returning; no intruder could have hidden himself in the house between the murders without someone knowing—and not without dripping blood wherever he was hiding. Abby was considered the primary target, killed during a rage, and Andrew was then dispatched to avoid his accusations. Lizzie had motive and exclusive opportunity, had even hinted at the possibility that something would happen to her father just the night before, and exhibited guilt in some of her remarks, inconsistencies, and actions. Basically, a circumstantial case against Lizzie was developed without the definitive identification of a murder weapon or physical evidence. Also, the case was hampered by the inability of the investigators to produce a corroborated demonstration of time and opportunity for the murders. They were not allowed to say that she had attempted to purchase prussic acid and could not introduce the significant inconsistencies in her inquest statements. In addition, they accepted without question the explanation of blood on garments and towels as menstrual blood.

In fact, a great deal of testimony about blood came in to the trial, from counting and interpreting the many blood spatters to using an analysis of the different conditions of the blood on the two victims to indicate how long apart they had been attacked.

Most damning was evidence that the dress turned over to the police by Lizzie as the one she had worn on the morning of the murders was not the same dress. Coupled with testimony from Alice Russell that she was with Lizzie after the murders when she burned a dress stained with "brown paint," this information made things look desperate for the defendant, although Lizzie's dressmaker stated that when she had made the dress the house was being painted. However, since the dress had been made only a few months earlier, Lizzie's claim that it was faded made little sense.

Crucial to the case was the presentation of evidence that supplied a motive for the murders. Prosecutors called witnesses to establish that Mr. Borden had intended to write a new will. An old will was never found. The new one would leave Emma and Lizzie each a paltry $25,000, with the remainder of Mr. Borden's half-million-dollar estate going to Abby.

The expensive defense team, led by Andrew Jennings, called witnesses to describe a mysterious young man in the vicinity of the Borden home that morning and Emma Borden to verify the absence of a motive for Lizzie. Emma even stated that she had urged Lizzie to burn the dress as per the family custom for useless garments. The team emphasized that no blood had been found on Lizzie, ignoring the testimony that it would not necessarily be the case that she would get blood on her from the way these murders were committed. (She could have covered herself with a garment and stood by the door frame in such a way as to reach in and slam the hatchet into Andrew's head. She might even have used Andrew's coat, and then rolled it and placed it under his head. No one checked the blood patterns on it to demonstrate that they were from wound leakage and not spatter.)

In the end, Lizzie won. The jury took just over an hour to find her not guilty. Relieved, she took her inheritance and moved to a finer house. The verdict is still controversial to this day.

The country barely had time to recover when the entire world was shocked by a devious man whose arrest brought attention to the criminal psychopath, as well as focus on a "real-life Sherlock Holmes." Ironically, the criminal had adopted that very surname as his own.

THE GAMESMAN

It was the discovery of a murder staged to resemble a suicide in October 1894 in Philadelphia that evolved into a case that eclipsed Jack the Ripper in its magnitude and invited criminal-mind specialists to test their theories. Marion Hedgepeth, a one-time cellmate of a man who went by the name H. M. Howard, informed the police about a scam. It had involved insuring a man named Benjamin Pitezel for $10,000 with the Fidelity Mutual Life Association in 1893 in Chicago, and then faking his death in a laboratory explosion by substituting a cadaver. The three parties to the fraud were then to share in the insurance payment, but Howard had run off with the money. Hedgepeth's letter alerted the company and they soon realized that H. M. Howard was actually H. H. Holmes (who was actually Herman Mudgett).

A company representative reexamined the circumstances of a body found badly burned at 1316 Callowhill Street in Philadelphia, which Holmes, accompanied by Benjamin Pitezel's young daughter, had identified as Pitezel. He had collected the money and left town with the girl and two of her four siblings. Company officers hired the Pinkerton National Detective Agency, and these agents gathered information about Holmes's numerous frauds in Chicago, which had provided him with funds to build a sinister three-story hotel not far from the grounds of the Chicago World's Exposition. On November 16, 1894, thanks to good leads, Holmes was arrested in Boston as he was preparing to leave the country. However, the Pitezel children were not with him.

On June 3, 1895, Holmes was tried for conspiracy to defraud an insurance company, and since the sentence was minimal, his attorneys advised him to plead guilty. Between that and the date for sentencing, reporters pressed for information about the Pitezel children, as did their mother, Carrie Pitezel. Detective Frank Geyer went on a highly publicized expedition to find them, and later penned a book about his painstaking trek. While Holmes had identified himself with Conan Doyle's famous fictional detective, Geyer would win renown as a real-life intellectual sleuth.

Geyer's wife and daughter had died in a recent fire, so his loss weighed heavily as he searched for children who were possibly dead. Holmes had said he had left them with a guardian, Minnie Williams, who took them to England. Geyer would later write, “Holmes is greatly given to lying with a sort of florid ornamentation.” The man, he believed, was an accomplished con artist, so his words could not be trusted (especially in light of the fact that Minnie Williams was also missing, along with her younger sister, Nettie).

Yet Holmes did admit to having had Alice Pitezel, fifteen, in his custody and to picking up Howard, eight, and Nellie, eleven. Alice and Nellie had written letters to their mother documenting their daily journey with Holmes, letters that he had never mailed and which were found in his possession upon his arrest. Geyer found no trace of Minnie Williams or the children where Holmes had said she would be, and the street name in London that Holmes had offered did not exist. Instead of going to England, where the clever offender was trying to direct him, Geyer focused closer to home.

On June 26, he set out by train into the Midwest, with Alice’s and Nellie’s letters to orient him, along with photos of Holmes and of the children, as well as an inventory of items and clothing associated with them. The possibility of finding evidence was minimal, yet the insurance company had readily provided funds for the trip. In Cincinnati, Ohio, Geyer showed photographs and asked around in various hotels for anyone who might have seen Holmes or the children, and he finally found someone who remembered the small group of travelers under the alias Alex E. Cook—a name Holmes had used in business matters before. That clerk pointed Geyer to a different neighborhood and through much questioning, he came across a woman who had seen Holmes and a boy together in a house to which a large stove had been delivered. But Holmes had then given her the stove. Geyer now felt that he "had firm hold of the end of a string that was to lead me ultimately to the consummation of my difficult mission." He went from there to Indianapolis, Indiana, Holmes's next known destination.

Here, Geyer found a trail that clarified Holmes's inexplicable game: He was moving his wife (one of three, all of whom were oblivious to the others) and the three children about in the same city without any party being aware of the others. Geyer could not understand why, if Holmes intended to kill the children, he would go to such effort and expense to move them so often. The puzzle deepened.

Geyer then went to Chicago and Detroit, the town from which Alice had written the last letter to her mother. He also learned that Holmes had added a third group to his game—Mrs. Carrie Pitezel and her other two children. He had placed her three blocks from where he roomed the three children in his care, but had not allowed them to realize it or see one another. Geyer spotted Holmes's pattern: He played games with people for his own satisfaction, adjusting his strategy to whatever seemed necessary to move them around. But Alice wrote something from that location to her mother that made Geyer pause with concern: "Howard is not with us now."

On a tip, Geyer went to Toronto and looked up real estate agents to find out if a man had rented a house for only a few days. "It took considerable time to impress each agent with the importance of making a careful search for us." He found a house that Holmes had rented, surrounded by a six-foot fence. Geyer struck out there, as the renter turned out to have been another man. Yet the intrepid detective felt certain the children had been killed somewhere in that town, so he persisted and found another rental that seemed suspicious. He learned that a man with children had asked for the loan of a spade to plant potatoes in the cellar and had brought only a bed, mattress, and large trunk to the house. A woman identified Holmes as the man who had rented the house from a photograph that Geyer carried. Geyer discovered that the house had a dark cellar accessible via trap door, and found an area of soft dirt. When he pushed a shovel into it, a stench arose and he knew his long, dark journey had produced what he had feared: human remains. After digging three feet, he found a small arm bone, so he employed an undertaker to take charge. In short order, they exhumed the corpses of two unclothed girls, which they believed to be Nellie and Alice Pitezel.

"Alice was found lying on her side with her hand to the west," Geyer wrote. "Nellie was found lying on her face, with her head to the south, her plaited hair hanging neatly down her back." A crew of men transferred them to a pair of coffins. "Thus it was proved," Geyer later wrote, "that little children cannot be murdered in this day and generation, beyond the possibility of discovery."

Searchers found a toy in the house that was listed in Carrie Pitezel's inventory of things that her children had owned, which assisted Geyer with a firm identification of the remains, as did pieces of partially burnt clothing. Carrie Pitezel, brought to Toronto, confirmed their identities.

But Geyer still knew they had yet to find Howard. He used logic and items mentioned in the letters to determine that Howard had been separated from the girls prior to their arrival in Detroit, so it was time to return to Indianapolis. He arrived on July 24 and proceeded with queries to real estate agents about short-term rentals from the previous October. By this time, Geyer's trek had drawn national interest and the Indianapolis newspapers heralded his arrival as a real-life Sherlock Holmes, which proved to be both a curse and a boon. He received a great many leads, but most of them came from people attempting to associate themselves with the famous quest, and they wasted his time. "Days came and passed," he wrote, "but I continued to be as much in the dark as ever." Geyer feared that "the bold and clever criminal" might have bested him and he despaired of finding the last child. His own grief spurred him onward.

In Philadelphia, Holmes kept track of Geyer's journey in the *Philadelphia Inquirer*. Initially he felt empowered at Geyer's lack of success, but with the detective's discovery of the girls' remains, Holmes was forced to devise a way to blame others. Even as he did so, investigators were analyzing the children's letters, and they sent ideas to Geyer based on things that had been previously overlooked. Geyer realized that the children had been in Indianapolis four days longer than investigators had believed, so he narrowed to two days the time frame not accounted for. He heard about the skeleton of a child found in Chicago and went to check, but it was not the young Pitezel. Instinct told him to stay in Indianapolis. "No less than nine hundred supposed clues were run out," he later wrote, but he persisted.

Geyer went to several outlying towns in the area, going through them as systematically as he had done in Indianapolis. In Irvington, he struck pay dirt. A man who had rented a house in October remembered Holmes because he'd been so rude, and he'd had a boy with him. Certain that he was at the trail's end, Geyer went to the rental property. However, he could detect no disturbance in the cellar's dirt floor, which discouraged him, but he found a trunk in a small alcove, and near it some disturbed dirt. Geyer dug into the area but turned up nothing. In a barn, he came across a coal stove, stained with a substance that resembled dried blood. He telegrammed Philadelphia with a description of the trunk and Carrie Pitezel identified it as hers.

A doctor poking around in a chimney at this place showed Geyer pieces of a charred bone—part of a skull and a femur—that he believed belonged to a male child. Geyer dismantled the chimney and found a set of teeth and a piece of jaw, identified by a dentist as being from a boy seven to ten years old. "At the bottom of the chimney," Geyer recorded, "was found quite a large charred mass, which upon being cut, disclosed a portion of the stomach, liver, and spleen, baked quite hard. The pelvis of the body was also found." Holmes had killed Howard here and incinerated him in the stove.

It was now August 27, fully two months after Geyer had left Philadelphia, but he now had everything he needed to prove Holmes a multiple murderer. It wasn't necessary, however, as Holmes went to trial for the murder of Benjamin Pitezel, wrote a confession in which he admitted to more than two dozen murders (although he later recanted), and his castle in Chicago had received a great deal of attention as a place of torture and murder. Found guilty of killing Pitezel, he was hanged. While he is remembered as one of America's more fiendish murderers, few people know the name of Detective Frank Geyer. Nevertheless, his investigative feat was heralded in his day as one of the stunning achievements of modern law enforcement.

THE STUDY OF MAN

European culture dominated many parts of the world, supported by Darwinian ideas that superior nations ought to rule inferior ones. The European white male was the standard against which other races were measured. While scientists proclaimed the neutrality of their methods, human bias nevertheless corrupted interpretations, filtering into legal proceedings. Alphonse Bertillon was among those infected. He played a key role in the 1894 court-martial of Captain Alfred Dreyfus, falsely accused of spying for Germany against France. At first, Bertillon indicated that the handwriting on a suspect document was unlike that of Dreyfus, but in an effort to win favor for authorities who clearly wanted to put Dreyfus away, he changed his opinion. He claimed that he simply hadn't spotted the way Dreyfus had disguised his handwriting during the interrogation in which he'd been forced to supply samples. The result was that Dreyfus went to prison, while a military cover-up protected the real perpetrator. The public was not duped, however, and the shameful affair divided France. It also triggered widespread scorn for the supposed science of handwriting analysis.

In another arena, Swiss anatomist Wilhelm His undertook a controlled study of the relationship of the skull to the face in 1895. He had acquired a skull that many believed was that of the late composer Johann Sebastian Bach, but he dared not experiment with such a precious item. Instead, he used data from working on twenty-eight cadavers, measuring the thickness of the soft tissue, to finally sculpt a likeness from the "Bach" skull. It turned out to resemble the composer closely enough to confirm that His was in possession of the right skull. To find out the general depth of the skin and muscles over the skull, His had plunged oiled needles into the corpses's faces and then attached a cork to the needles. Once the needle hit bone, the cork rested at the skin's surface. He then pulled the needles out, measured them, and made drawings based on the measurements from different areas. Thus, he devised a sort of topographical map, leaving a legacy for future anthropologists and forensic sculptors.

Around this time, Lombroso came back into the picture with a new invention, the sphygmograph, which measured changes in blood pressure and pulse during the interrogation of suspects. The person donned an airtight volumetric glove attached to a thin piece of rubber that stimulated a pen to record the variations evident in the suspect's blood flow. Lombroso assumed that when the person lied, his or her blood pressure would change, and when compared to what was said during that time frame, it could be established as truth or a lie. Thus, he provided the first lie detector, based on physiological changes.

Wilhelm Conrad Röntgen discovered x-rays that year and founded the science of x-ray crystallography, wherein refracting x-rays through a crystal diffracted them, allowing them to be caught on a photograph. While not yet applied to criminal investigations, hindsight would eventually note this discovery.

In addition, odontology got another boost, albeit from an enormous tragedy in Paris in 1897. Under France's Third Republic, the first charity sales lured crowds of aristocrats to come to stalls to purchase goods, thereby supporting worthy causes. During one such bazaar on May 4, in front of the wooden building where more than one hundred charities were represented, a gas lamp exploded in a cinematograph. The fire spread quickly, and even before the fire brigades arrived, the building had collapsed. While many survivors were pulled out, 140 had been killed, which left officials with a massive number of identifications to make among weeping and frightened relatives of potential victims. The bodies were taken to the nearby Palais of Industry and laid out in rows. Relatives managed to identify most of them, taking them away for burial, but thirty bodies were so badly burned they remained where they were. Among those who came to find a loved one was the Paraguayan consul, looking for the body of Duchesse d'Alencon, sister to the empress Elizabeth of Austria. Observing that the teeth of these victims remained intact, he called on dentists who may have treated the victims to attempt to use their records to identify them, and with this approach most were returned to their families. The reports about these dental identifications, while not forensic in the legal sense, were preserved for future students of dentistry and forensic odontology.

The following year in 1897, Adolph Luetgert, a sausage maker in Chicago, Illinois, was arrested when his second wife, Louisa, disappeared. It turned out that Luetgert was a notorious seducer, as well as an abuser, and had allegedly told one mistress he wished his wife were dead. Since he had ordered a large quantity of caustic potash two months before Louisa vanished in order to "make soap," and since she was last seen in his company on May 1 outside the sausage factory, he was a viable suspect. Reportedly, on May 2, the night watchman had noticed a sticky substance on the floor, mixed with bone fragments, and a few days later Luetgert began giving out the story that his wife had simply walked away.

The police learned about the substance on the floor, searched the factory, and discovered a vat full of suspicious brown sludge. They drained it and found bone fragments and two rings, one of which bore the initials, *L. L.* Luetgart was bound over for trial. His defense attorneys presented witnesses who had seen Louisa in other places, but the prosecutors engaged the services of George Dorsey, an anthropologist at the Field Columbian Museum. He identified some of the bone fragments as human, including a thigh bone from a female. His presence at the October trial marked the first time a forensic anthropologist had testified in court. In addition, a physician and a professor testified that the sludge from the vat was identifiable as dissolved human flesh.

The jury hung, but the second trial in February 1898 resulted in a conviction and Luetgert received a life sentence. Although there was no body, evidence and circumstances supported the possibility that he had boiled his wife in the vat of caustic potash. However, rumors flew that he'd actually ground her up into sausage.

Then in 1898 in the area of ballistics, chemist Paul Jeserich got involved in a murder case in Germany when the police asked him to determine if a bullet removed from the victim had been fired from the suspect's gun. Rather than make an examination through observation, as Lacassagne had done, he opted for a different kind of analysis. He fired a bullet from the gun in question, and then took photographs of the test bullet and the one removed from the victim, which amplified their qualities. Looking at both, he identified specific markings that appeared to link both projectiles to the gun. The suspect was convicted and sent to prison. But it would be fifteen years before Victor Balthazard would note that each gun placed a distinctive mark on the bullets it fired.

Then fingerprints emerged again. This time, in the

effort to win widespread acceptance in the law enforcement community, the technique got the right attention. Just before the turn of the nineteenth century, Francis Galton ensured that Edward Henry could present a paper at the meeting of the British Association for the Advancement of Science. But contrary to his expectation, Henry bypassed mention of Galton's system and instead discussed his own ideas, apparently devised from a man named Azizul Haque, who had worked for him in India. Henry and Haque used ridge counting to sort patterns into finer classifications for more precise identification than Galton's three-pattern typology allowed. Afterward, officials in Britain convened a committee to consider whether British law enforcement should adopt this new method. However, Galton attacked Henry in a professional journal, indicating errors with Henry's system and challenging him to prove himself with a large collection of tested prints.

Henry obliged, offering a trunk full of fingerprint lifts that revealed, one after another, that individual fingerprints were unique and that his classification system worked. He published a book, *Classification and the Use of Finger Prints*, saying that his ideas were already in widespread use in India. The British committee decided to give Henry's method a cautious try, so he received the post of assistant commissioner of Scotland Yard. In 1901, he established its Fingerprint Branch, as well as taking over the Criminal Investigation Division (CID).

That year, *bertillonage*, still keeping its ground, identified just over five hundred suspects as repeat offenders, while the next year fingerprinting proved to be more than three times as successful. Since it was an easier and more practical method, and appeared to be reliable, Scotland Yard took steps to make it the method of choice. Still, Henry had a job ahead of him convincing the police force out on the crime beat that this was a useful system. He needed a big case. In the meantime, handwriting analysis was in fashion again, at least in another sensational trial in New York.

A rivalry had developed in the Knickerbocker Athletic Club between Harry Cornish and Roland Molineaux, a chemist. Just before Christmas in 1898, Cornish received in the mail an anonymous gift of apparent bromoseltzer, placed inside a silver container shaped like a candlestick. Thinking it a holiday joke, he offered it to his aunt, Katherine Adams, for a headache remedy, but when she mixed and drank it, she convulsed in agony and died. The autopsy confirmed that she had ingested cyanide.

Molineaux, thirty-one, was a chemist and the superintendent of Morris Herrmann & Company. The plant had a laboratory containing chemicals from which poison, including cyanide of mercury, could be produced. Indeed, Molineaux had been a suspect in 1897 when his rival for the affection of a woman had also died from a mailed poison. He was arrested and tried. The prosecutor included statements from eighteen expert witnesses on handwriting analysis, including two eminent analysts, John F. Tyrrell and Albert S. Osborn, while the defense attorney used only one authority on questioned documents, David Carvaho. The prosecution's case seemed stronger, as the experts stated that Molineaux's handwriting matched that found on the package containing the anonymous gift, and they even recognized his efforts to disguise it. They based their conclusions in part on the recurrence of the same misspelled words. Molineaux claimed to be entirely innocent, but the judge allowed testimony about the 1897 case, so the jury found him guilty.

Molineaux appealed the decision. When the appellate court found that information about a murder for which he'd never been arrested or convicted should not have been included in his original proceedings, he received a new trial. This time, fewer samples of his handwriting were used and his own attorney hired experts who stated that one could not distinguish the aspects of handwriting that the original prosecution's experts had described. This jury wasted no time in finding Molineaux not guilty.

Professionals all over Britain and America were now debating the issue of experts in the courtroom, which had become an increasingly lucrative arena. People could now sell their expertise and many "experts" claimed to have scientific backbone in their techniques when they didn't. Professionals contradicted one another on supposedly "certain" results, and accusations flew about each other's lack of moral fiber or poor handling of evidence. It was easy to point to cases throughout the century in which people had been harmed by so-called scientific research or interpretation, and it seemed that the courts now encouraged more of the same. However, juries did not trust it as yet, especially since they had no means for deciding which expert was right, and even authentic experts had sometimes failed to deliver what they'd promised.

Even so, scientists continued to improve their techniques and to hope for courtrooms to tighten up the standards. They suggested alternate tribunals for selecting courtroom experts and reexamined their epistemological and ethical assumptions, knowing that they could not continue to present excuses for widely differing opinions on the same evidence. Courtroom stipulations about expertise were certainly in the offing, albeit several decades away. For now, the scientific community had to police itself, but as long as there was money to be made, there would be charlatans with their junk science, as well as good science presented poorly in the adversarial arena. More cases were coming that would place these problems into a glaring spotlight.

The nineteenth century ended with a great many forensic technologies and inventions coming into their own. Fingerprinting would soon win its place as one of the most important discoveries for law enforcement, and the new century would usher in the age of identification. So much more was in store, including more prominent scientists and investigators who would gain their own renown for what they contributed to the balance of justice.

SEVEN

SHERLOCK'S LABS

IN THE RED

The early 1900s ushered in many discoveries in forensic science, from a wide variety of fields. While even Sherlock Holmes pondered the analysis of blood, echoing a significant concern for law enforcement, the best test thus far was the microscopic examination of stains that could distinguish blood from substances such as juice or dye, as long as they were sufficiently fresh to show the blood's corpuscular structure. One doctor in France claimed that heat would make blood give off a certain odor, but his "test" never showed up in a court of law. While the spectroscope from as far back as 1859 could assist in detecting hemoglobin, animal blood could not be distinguished as yet from human.

In 1900, Austrian immunologist Karl Landsteiner, a professor at the Institute of Pathology and Anatomy in

Vienna, named and standardized the different blood groups, based on the structure of red blood cells. He found that they carried antigens, which produce antibodies to fight infection, and different people showed different conditions. In a centrifuge, Landsteiner placed blood from colleagues who had graciously volunteered, separating the red blood cells from the watery serum in which they were carried. Then he added red blood cells from other subjects and identified two distinct reactions—clumping and repelling. He labeled them type A, if the antigen A was present with the anti-B antibody and no antigen B. Those with the antigen B but no antigen A became type B. A third reaction in which both antigens were absent was labeled C (later relabeled as O). Two years later, yet another type of serum showed up with both antigens present, so it was labeled type AB. So now the police could at least *exclude* suspects whose blood type did not match blood at a crime scene.

Around the same time, another important serological discovery occurred. German biologist Paul Uhlenhuth, from the Institute of Hygiene in Griefswald, was studying ways to develop a serum to fight hoof-and-mouth disease. He learned from experiments that if he injected protein from a chicken egg into a rabbit, and then mixed serum from the rabbit with egg white, the egg proteins separated from the liquid to form a cloudy substance: an antibody. He called this substance precipitin. As he proceeded, he found that the blood of each animal had its own distinct protein, and then after injecting human cells into the rabbit, he realized that the test was also applicable to humans. That was big news for law enforcement because crime suspects often claimed that blood on their clothing was from animals, and to that point, their stories could not be disputed with proof. With the precipitin test, those days appeared to be over. To be certain about this result, a coroner asked Uhlenhuth to test some dried bloodstains from both animals and humans, and the results proved the test to be reliable.

Just four months after Uhlenhuth announced his discovery, a particularly brutal crime brought the test into the forensic spotlight. In the German village of Göhren, near the island of Rügen, two young boys, Peter and Hermann Stubbe, failed to come home after playing out in the woods. Villagers assisted in the search, even into the night. As the sun rose, several body parts that appeared to be from children were discovered in the bushes. Then the searchers came across the bodies, with their limbs and organs removed and strewn about. Gruesomely, Hermann's heart was gone, and both boys had been bludgeoned with a stone.

The police learned that people had seen the boys talking with a fruit seller named Ludwig Tessnow from Baabe, so they went to question him. Although he denied any involvement in their murders, a search of his home turned up laundered clothing that bore suspicious stains. Tessnow claimed that they were from wood dye, which he used for carpentry. There was nothing more the police could do since they had no evidence, but then a local magistrate, Johann-Klaus Schmidt, recalled a similar crime not far away—and involving the same suspect.

Three years earlier in Osnabruck, Germany, two young girls had been found in the woods, butchered and disemboweled. The man seen loitering near the woods, his clothing stained, was Tessnow. At that time as well he had claimed that the stains were from wood dye. The circumstances now looked much more suspicious, especially after the prosecutor, Ernst Hubschmann, heard that a farmer had identified Tessnow as the man he had seen near his field just before finding seven of his sheep slaughtered and ripped apart. Still, no one had actually seen the suspect perform such brutal crimes on human or animal, and the prosecutor needed more than circumstantial associations. Hubschmann came across Uhlenhuth's paper "A Method for the Investigation of Different Types of Blood," so he asked Uhlenhuth to examine Tessnow's clothing. Uhlenhuth agreed, applying his method to more than one hundred spots. While he did find the presence of wood dye, he also detected traces of both sheep and human blood. With this evidence and the surrounding circumstances, Tessnow was tried, convicted, and executed.

Then in 1902, a murderer in France who said that blood on his clothing came from skinning a rabbit was also convicted via the bloodstain test, when the blood proved to be human. Thanks to a scientist, law enforcement had gained an impressive new tool.

∴

MAKING MATCHES

The first time an expert proved in court that a specific gun was used for a murder was in America in 1902. Oliver Wendell Holmes, a judge, liked to keep up with scientific innovations and he came across a book about firearms identification. When a case came up that provided an opportunity to use a scientific demonstration in court, he called a gunsmith to test-fire the alleged murder weapon into a wad of cotton wool. With a magnifying lens, he showed the jury that the marks on the bullet from the victim matched the one fired from the suspect's gun.

During that era, *bertillonage* still dominated human identification, but detectives working with fingerprint evidence continued to seek acceptance for their methods, which they believed were much more manageable, scientific, and accurate than taking body measurements. In England, Edward Henry, who had three good investigators working with him in the Fingerprint Branch, finally found the case he needed. It was June, time for the Derby in Epsom, which drew plenty of spectators, along with thieves and pickpockets. Assistant Commissioner Melville Macnaghten from Scotland Yard's CID suggested that Henry take fingerprints of thieves at the Derby. There would be no end of opportunities to prove the method and it would certainly get publicity.

That day, the police arrested more than fifty pickpockets, taking impressions of their fingerprints. Two officers worked through the night to identify repeat offenders and they were able to show that more than half of these men had been arrested before. Thus, they received a harsher sentence than they had expected. One man actually said, "Bless the fingerprints. I knew they'd do me in."

The next case, two months later, involved a home invasion, in which a set of fingerprints was lifted from where the offender had touched wet paint. From a list of burglars who were required to give their prints, the investigator was able to match one from the paint to Harry Jackson. Edward Henry was determined to take Jackson to court using this single fingerprint as evidence. A prosecutor of stature, Richard Muir, accepted the case for the sole purpose of bringing attention to this technique. During the proceedings he told the jury that fingerprint evidence had never before been used in a jury trial in an English court. He had investigator Charles Collins explain the methodology in detail, which the jury found fascinating. Collins offered a photographic enlargement of both the print from the paint and the inked impression from Jackson, describing a fingerprint as a calling card. Although Jackson claimed he had nothing to do with the crime, the jury convicted him and gave him seven years in prison. A few newspapers picked up the story, but there was immediate backlash from professionals, including Henry Faulds, who claimed that a method that relied on a single print had not been proven. There was another setback when a clerical error at the Fingerprint Branch forced the detectives to admit they had made a mistaken identification. Thus, while it got attention, this new approach failed to have an immediate positive influence on public acceptance.

But in the United States, an incident at Leavenworth Penitentiary in Kansas in 1903 convinced law enforcement of the technique's superiority. A convict named Will West came in for processing. An agent located his card on file, but West protested that he had never been there before, so they could not possibly have his card. The guards took his body and head measurements, which were close enough to those on the card to believe that West was lying, but again he was adamant. Looking into the matter, the guards found another William West in the prison, who bore a strong resemblance to the new man coming in. This unique coincidence proved to be a blow to anthropometry, but a real gain for fingerprinting, since that, at least, distinguished the two men.

Another case in Britain also helped to give science a boost, but only after it destroyed a man's life. Back in 1895, a woman had identified Adolf Beck, owner of a copper mine, as the thief who had stolen her watch. Soon ten other swindled women nailed him as well, so Beck was convicted and sentenced to seven years in prison. After he got out, he was again identified as the person who had committed a theft and nearly returned to prison a second time, but then John Smith was arrested and he confessed to all of the cons. His physical similarity to Beck was astounding, highlighting a key problem with eyewitness memory: It could easily be mistaken, especially where resemblances were uncanny. While Beck was pardoned, he had lost his business and his sense of trust. Inspector Melville Macnaghten had witnessed this case and it fueled his enthusiasm for utilizing fingerprints. Had such evidence been available in 1895, he knew, Beck would have been spared his ordeal.

British investigators were soon to find a much more sensational, albeit still tenuous, case to demonstrate their methods. It was early in the morning in Deptford, England, on March 27, 1905, when a young man entered Chapman's Oil and Colour Shop on High Street and found owner Thomas Farrow bludgeoned to death under an overturned chair. The police arrived and found Farrow's wife upstairs, also assaulted and in need of immediate medical attention.

An empty cashbox revealed robbery as the motive. Chief Inspector Frederick Fox and Assistant Commissioner Melville Macnaghten deduced from that that Farrow had been duped into opening the door. The robbers had then gone up to the bedroom, bludgeoned Mrs. Farrow, found what they were after in the cashbox, and fled. Yet they'd left behind their masks.

Macnaghten hoped for a good fingerprint analysis so he instructed his officers to process the place carefully. With a handkerchief, he picked up the cashbox, saw what he believed was a print, and carefully made an impression for the lab. It was indeed a print and appeared to be from a thumb. Yet no prints on record from housebreakers provided a match. Witnesses had seen brothers Alfred and Albert Sratton in the area, so the inspectors took their prints. After hours of waiting, a match was made with the elder brother. This evidence was prepared for court, along with witness identifications, and agencies around the world awaited the results.

Richard Muir took the case, although he realized that with lives at stake, the jury might resist convicting on this new type of evidence. Much hung in the balance for the fingerprinting technique. If the print was barred from court, that would be a considerable setback. If admitted and the evidence actually contributed to a conviction, it would become a legal precedent, with international reverberations.

But then Henry Faulds once again became a vocal detractor. Stung by the lack of recognition he'd received, he insisted on the necessity of having all ten prints for an identification. Nevertheless, Scotland Yard proceeded with the case.

At first, the trial faltered. The defense attorney vigorously disputed each prosecution witness, yet on the other hand the brothers had masks in their possession similar to those found at the crime scene and they had tried to persuade someone to give them an alibi. As well, they were richer directly after the murder and they had changed their appearance.

Finally, the fingerprint experts gave their testimony. When Charles Collins showed the jury enlarged photographs illustrating how the thumbprint from the scene matched the elder Stratton on eleven points of comparison, he spoke with confidence. He told them that he'd worked for more than four years with files that numbered over ninety thousand prints. Although defense expert Dr.

John Garson pointed out dissimilarities between the prints, Collins rebounded with an explanation: He said these differences were the result of different types of pressure. He proved his point by taking prints from members of the jury, with different degrees of pressure, to show them differences in their own prints. It was an excellent visual demonstration, since they could see for themselves that prints they knew to be their own could have slightly different appearances. Further undermining the defense expert, the prosecution produced a letter he'd written to the effect that he would offer testimony to the highest bidder, so he was dismissed.

It took the jury two hours, but ultimately they accepted the fingerprint interpretation, convicting both men and sentencing them to hang, which went straight into the newspapers around the country, and was soon picked up internationally. That case opened the door for other police departments to trust the fingerprinting techniques.

That year, the New York State prison system began the first systematic use of fingerprints in the country for criminal identification, and by 1910, an appeals court would declare that fingerprint technology had a scientific basis.

PATTERN ANALYSIS

At this time, there were developments in other areas as well, some positive and some negative. In 1904 Oskar and Rudolph Adler developed a presumptive test for blood based on benzidine, a new chemical developed by Merck. But then around 1905, the science of wound pattern analysis came into doubt as expert debated expert in another sensational case. Jeanne Weber, a Frenchwoman, was accused of killing several children of acquaintances and relatives, but the government's pathologist, Dr. Leon Thoinot, repeatedly insisted that the deaths were accidental.

In January 1906, Weber appeared in court, where Thoinot described studies from the past eighty years concerning marks left by manual strangulation. In 1888, Dr. Langreuter had opened the skulls of fresh corpses and observed what occurred inside as his assistant choked the corpses or strangled them with cords. Langreuter had noted that victims of manual strangulation show specific bruises on the neck, and dotlike facial and muscle hemorrhages, called petechiae. Thoinot had to admit that his tests on the corpses based on these studies had come out negative. While circumstantial evidence and witness reports supported a finding of murder, forensic science could not support it. Weber was acquitted, but the case was not over.

Under another name, she became a governess and another child died. She was arrested again and subjected to Thoinot's analysis, but he rejected the cause of death as strangulation, so Weber received another reprieve. But then she was caught in 1908 attacking a young boy. The examining doctor made an exhaustive photographic documentation, knowing that Thoinot would evaluate the corpse. However, Thoinot avoided impugning his reputation and decided that this time Weber had acted out from the stress of her numerous arrests. He recommended that she be sent to an asylum. Those doctors who had lost ground in this case were determined in the future to avoid what they considered a travesty, so they set about to improve the system. They believed that careful analysis had proven the case, even if an imposing authority had swayed the jury. With more convincing proof, they could have saved the lives of at least some of Weber's victims.

During this time in Germany, a young woman named Margarethe Filbert was murdered. She had disappeared on May 28, 1908, and her headless corpse was located the next day, with hairs clutched in her fist. Although she was provocatively posed, there was no evidence of sexual assault. District Attorney Sohn, also the chief investigator, had read an article celebrating chemist Georg Popp as a modern-day Sherlock Holmes, a man who used microscope analysis to solve crimes. He called on Popp in Frankfurt, Germany, to request his assistance.

Popp had become fascinated with the application of chemistry to forensic analysis after a case he had worked in which he had analyzed spots on a suspect's trousers. He had read Hans Gross's book *Criminal Investigation,* and had even managed to identify a thief in his own lab by using vapors to expose a latent fingerprint. He continued to assist the police with fingerprint photography and the identification of fibers and soil traces.

In the case of the headless corpse, Popp analyzed the hairs and said that they were from a woman, and were possibly Filbert's, but he needed the entire head to be certain. A farmer and local bully named Andreas Schlicher came under suspicion, and scorings from under his fingernails were analyzed with the Uhlenhuth test, yielding traces of human blood. Popp requested the man's clothing for microscopic analysis, but curiously, Sohn refused to send them. Despite the existence of methods that would clearly help, the local authorities apparently balked at taking this approach too far.

Then another detective took over the case and he wanted to see it resolved, so he sent the clothing. Popp found evidence of blood on the man's shirt and trousers, with obvious attempts to wash it off. He also examined the suspect's shoes, finding that they bore several layers of soil with embedded purple and brown fibers. Some of the soil was similar to soil from the crime scene, but not with soil found in other places the suspect claimed to have been. Popp used a spectrophotometer, based on the spectroscope invented in 1859 by Robert Bunsen, because he wanted to compare the spectrum of emission lines that came from the dyes in the fibers from the shoes and the victim's clothing. He found that the purple and brown fibers were identical in color and consistency to the victim's skirt.

In the first documented case to focus on the analysis of soil and the chemical composition of fiber, a jury found Schlicher guilty, based largely on this impressive physical evidence, and he then admitted his deed as a crime of opportunity. He had hoped to rob the woman, he said, but when it turned out she had no money, he had removed her head as a gesture of anger. He led authorities to where he had placed it.

Popp was not the only one to espouse the use of the microscope. Lacassagne did as well, when he taught its merits to students at the Lyon Institute of Forensic Medicine. One of them, Émile Villebrun, went on to study the value of fingernails in forensic investigation. Not only could evidence be found under fingernails, he indicated, but they also make distinctive marks when scraping human flesh—sometimes perpetrators to victims, and sometimes victims to perpetrators.

MURDER SQUAD

Scotland Yard's most elite department of detectives was founded in 1907. At the time, there were many unsolved murders around England, so home secretary Herbert Gladstone decided to create an elite unit from members of the Criminal Investigation Department. London had plenty of detectives—around seven hundred—but the provincial cities were lacking. London officials found that when cases with special circumstances arose, the local law enforcement generally muddled them. They might eventually request a consultation from Scotland Yard's more experienced force, but usually after it was too late for the detectives to offer much help. In addition, there were turf wars, even when help was requested. The home secretary decided that since he had men with considerable experience "with a particular class of people" he thought their talents should be better utilized. He moved to give Scotland Yard greater jurisdiction over serious crimes in locations outside London. For that, he designated a small number of talented men for this task. They would come to be known as the Murder Squad, although they received no such official title. Gladstone set to work advising local chief constables of the experience these men had and inducing them to take advantage of their experience when needed.

The first chief of the Murder Squad was Frank Froest and he set about making the group of four men under his supervision into a unit in which they would serve with pride. They received more training and assisted one another, soon developing an identity among law enforcement, as well as working on crimes that garnered media attention. In an age when Sherlock Holmes stories were all the rage, these men gained national prominence.

However, it proved difficult at times to investigate murders some distance from London, because people with something to hide often corrupted crime scenes and removed clues. In one case, a young amputee saving money for a cork leg was murdered. His mother, Flora Haskell, said that half the money was missing. She also described a running man who had thrown a knife at her and spattered her clothing with blood. That's when she'd found her son dead in his bed. A story like this sounded suspicious to seasoned detectives, and it was, since no one else in the village reported seeing a stranger of the description she gave. Given that the boy, age twelve, was probably a burden, and that she had needed money, it seemed clear that she was the likely culprit—especially when she admitted to having cleaned up all the blood. Indeed, the local doctor had even washed the body by the time the detectives arrived, and they were on the scene in less than twenty-four hours. They soon learned that the knife used belonged to the family's household, and recently had been sharpened. The circumstances added up against Flora, but her counsel managed to convince the jury that the lack of blood at the scene meant the key piece of evidence had been unavailable. They acquitted the woman, believing the case against her had not been proven.

But the police believed otherwise. The Haskell case inspired the home office to notify the authorities in all localities that junior officers were to be instructed that all such crime scenes must be preserved in the condition they were found, and a guard should be posted until experienced personnel could arrive. As these safeguards were instituted, one member of the Squad, Walter Dew, won widespread fame for his handling of a missing persons case. That same incident brought a new expert into the courtroom who would go on to become one of Britain's premiere forensic pathologists.

Dr. Bernard Spilsbury already had made a name for himself as a medical practitioner during the early 1900s at St. Mary's Hospital in London, associating him firmly with the coroner's court, where he performed "necropsies" to determine the cause of death in civil and criminal cases. In 1910, he examined two cases of medical malpractice, but then a physician came to Spilsbury's attention who had committed murder outside the practice.

The suspect was American-born Hawley Harvey Crippen, a patent salesman as well as a physician. In January, he boarded the S.S. *Montrose* bound for Quebec with a young boy, who was actually his disguised mistress, Ethel Le Neve. A telegraph to the ship (a first for radio technology) effectively stopped him, and he was arrested on board and returned to England for trial in the murder of his wife, Cora.

In fact, she had been his second wife, and they had moved to England together in 1900. Cora fancied herself an opera singer and at some point she apparently got involved with a performer named Bruce Miller. Around the same time, Crippen hired Le Neve as a typist. They remained acquaintances for five years before becoming lovers.

The Crippens resided at 30 Hilldrop Crescent, generally living above their means. Crippen, forty-eight, cut an insignificant figure, being small, bespectacled, and mild-mannered. People viewed him as kindly. One day, Cora simply vanished without saying good-bye to a single friend. She had last been seen at a dinner party on January 31, and within a day, Crippen had pawned some of her jewelry. He also sent letters in her name to various organizations offering Cora's resignation, and he moved his mistress into his home as his new companion.

Le Neve took over the household and hired a French maid. She was even seen by people who knew Cora wearing some of Cora's furs and jewelry. Crippen told friends that Cora had gone to visit friends in America, and very soon he was adding that she had grown ill and died there. But his public flaunting of Ms. Le Neve made his story suspicious. A friend of Cora's tried to get answers and when he could not, he went to Scotland Yard. Questioned by Walter Dew from the Murder Squad, Crippen said that the embarrassing truth was that Cora had run off to be with a lover, Bruce Miller, and they were together in Chicago. Nevertheless, the officer searched Crippen's house but observed nothing to indicate foul play.

But people who knew the Crippens were aware of Cora's domineering behavior and Crippen's obviously resentful restraint. Apparently she had learned about the mistress, because in December 1909, she threatened to take the money from their joint savings account and leave. In January, Crippen had ordered five grains of hydrobromide of hyoscin at the shop of Lewis and Burrows on New Oxford Street—a considerable amount that demanded a special order. He picked it up on January 19. Shortly thereafter, Cora "ran off."

The day after Dew had visited, Crippen used a pseudonym to board the S. S. *Montrose* in Antwerp. This left Crippen's house unattended, and since his sudden departure had again raised suspicions, a team of officers performed a more thorough search. This time, the results were quite different. Beneath some bricks in the coal cellar, Cora's dismembered and decomposing torso, sans some organs, bones, and the genitals, was found, and it was determined from tissue analysis that she had ingested a lethal dose of hydrobromide of hyoscin.

Even before these tests were performed, a police sketch artist made a drawing of the fugitive, based on old photos. But the *Montrose*'s captain had already recognized Crippen from photos of the "Cellar Murderer" in a newspaper he'd brought on board. He sent a telegram back to officials on shore regarding his suspicion. Crippen was detained, while a police officer boarded a faster ship to arrive before him in Canada. On July 31, 1910, Crippen was arrested and returned to England.

During his sensational murder trial, his counsel questioned that the remains were those of Crippen's wife but failed to produce evidence that she was alive beyond February 1, as well as to demonstrate why she would leave on a journey in February without her furs. A chemist testified that twelve days before Cora disappeared, he had sold the fatal drug to Crippen, who had never purchased it before. Then a doctor described the tests he had used to determine that Cora had been poisoned, while Spilsbury and another doctor identified a scar from the torso's lower abdomen as the result of a surgical procedure that Cora had endured, as well as proving the scar was not just a postmortem fold in the skin. Spilsbury even demonstrated this with a microscope. After five days of testimony, the jury took twenty-seven minutes to convict Crippen and he was hanged. (Ethel Le Neve was acquitted of any involvement and she sold her story to the press.)

Spilsbury was soon testifying in yet another fatal poisoning. In 1911, insurance agent Frederick Henry Seddon was arrested for poisoning his lodger, Elizabeth

Barrow, to steal her assets. The circumstantial evidence was that Seddon had persuaded Ms. Barrow to transfer a considerable sum of stock funds and some property to him in exchange for a regular annuity. Not long afterward, she grew ill and died. Since she had suffered from asthma, her doctor decided the cause of death was heart failure. But then a relative announced to the police that her cash had disappeared, so her remains were exhumed.

Scotland Yard authorized Spilsbury to perform another autopsy and toxicologist William Willcox to make some tests. Spilsbury failed to find evidence of heart failure, so Willcox, who had noted arsenic in the organs, hoped to try to quantify it. He ran hundreds of weight tests for arsenic and then used the arsenic mirror method and the assumption that arsenic would distribute evenly in the body to figure out how much arsenic was in each of the poisoned woman's internal organs. He came up with a figure of two grains, which was sufficient to kill a person. Setting a precedent for forensic science, he calculated the amount via body weight in milligrams.

But defense attorney Edward Hall was ready for him. He noted that Barrow's weight had diminished considerably in the grave, from one hundred forty to around sixty pounds and suggested that this would exaggerate the arsenic concentration in her current state. Willcox, for all his brilliance, had not considered this possibility, and he was forced to concede that the dose of arsenic found in her might not have reached fatal proportions. In addition, the corpse showed arsenic levels in the hair that indicated a long-term administration, but Barrow had shown no symptoms. Could it be that she had absorbed the poison over time from the arsenic flypaper in her room?

Willcox then performed another experiment in which he soaked hair in fluid from the coffin, getting the same amount of arsenic as Barrow had. Thus, he was able to conclude that the source had been the fluid leaking from her body. In the end, his careful work stood up against vigorous challenge, and when Seddon showed just how greedy he was for money, his performance on the witness stand facilitated his conviction.

In 1914, there was yet another famous case that involved both Willcox and Spilsbury. Margaret Lloyd had died in her bath in Highgate, England. A relative of a victim of a similar drowning spotted her obituary and brought the matter to the police, who noted the criminal record of her husband, George Joseph Smith. Indeed, he had not only married Margaret Lloyd under an assumed name, but had married three times and each wife had drowned in her bath. Still, it seemed quite unlikely that someone could have drowned women in a bathtub without them struggling fiercely, and there had been no mark from violence on any of the bodies. The latest victim had indeed drowned, as evidenced by foam in her lungs, but there was no indication of force.

Spilsbury set up an experiment using young women in bathing outfits who agreed to sit in bathtubs and allow him and a detective to try to drown them. After repeated failures, it seemed impossible to make a person drown in this context. But then the detective deduced the answer: Smith had killed them by grabbing them by the feet and pulling them helplessly into the water. The quick action and rush of water made them helpless. In fact, one of the participants went unconscious at once. It seemed clear that Smith had figured out what to do and had done it three times. Then he enriched himself on their money or insurance. The "Brides in the Bath Killer" went to the executioner in 1915 shouting, "I am in terror!"

AMERICA'S ELITE

American president Theodore Roosevelt authorized a federal investigative agency so on July 26, 1908, Attorney General Charles J. Bonaparte ordered a force of investigators to report to the Department of Justice's (DOJ) Chief Examiner, Stanley Finch. They were to handle all investigative matters related to the DOJ except specific types of financial frauds. The force was named the Bureau of Investigation (BOI), and the agents busied themselves with such crimes as land fraud, interstate commerce, and involuntary slavery. By the end of the year, the BOI had thirty-four permanent agents.

In 1910, Albert S. Osborn, another American, published *Questioned Documents* to show the forensic value of document examination, although in this field, the court still questioned the subjective nature of even expert interpretation. Nevertheless, Osborn was becoming a noted expert in this area, while Victor Balthazard, a Parisian medical examiner, published the first significant study of hair, *Le poil de l'homme et des animaux*. The following year, U.S. physiologist Thomas Hunt Morgan and his students scrutinized fruitflies and demonstrated that chromosomes carry inherited information—a discovery that would one day have implications for crime investigation.

By now, another fictional detective had caught America's interest. Jacques Futrelle published "The Problem of Cell 13" in 1906 in which he introduced "the Thinking Machine," Professor Augustus S. F. X. Van Dusen, who spent most of his time in his lab thinking about difficult problems and who had been honored by a multitude of universities and scientific institutions for his cerebral skills. "He was a Ph.D., an LL.D., an F.R.S., an M.D., and a M.D.S." In a series of stories between 1906 and 1912, when Futrelle died on the *Titanic*, he impressed readers with the idea of how science, ingenuity, and logic can solve seemingly impossible dilemmas, including complex crimes. In the most famous story, Professor Van Dusen accepts a dare to be locked inside a cell without tools, to think his way out within a week. "I've done more asinine things than that to convince other men of less important truths," he says, and he succeeds. The fact that his success depends as much on luck and the availability of a rathole and other people as on his ability to think logically may escape the reader. That's probably due to Van Dusen's own dismissal of assistance as minimal. While he's certainly clever, one can easily devise a cell without the rat hole that would likely defy even his impressive wit. Throughout these stories, the professor was all brain and at the time his approach had nearly as great an influence on popular ideas about investigation as did Sherlock Holmes.

The point of a Thinking Machine was twofold: to be able to think one's way through any problem, no matter how impossible it may seem, and to prove the superiority of reason as a human faculty. The scientific enterprise emphasized an objective, reasoned and systematic approach to the problems of reality and how things worked. A good mind and sharp focus, along with the ability to bracket emotional attachment to ideas or results, was the hallmark of the laboratory scientist. The ideal was for them to pursue their work like machines. That this understanding minimized the passion that drove nineteenth-century scientists to apply their work to investigation apparently escaped those in the new century who hoped to make their mark in the field. In fact, the character Van Dusen was portrayed as a man much in demand internationally by institutions wishing to gain his sponsorship, because they wanted to align themselves with his form of rarified thinking: Fictionally speaking, his was the "foremost brain" in the sciences. But along with it came a "crabbed" personality, marked by disdain, arrogance, intolerance, and impatience.

CRIME LAB

Lacassagne and Villebrun both influenced another young Frenchman, who grew up in Lyon avidly reading translations of Hans Gross, as well as the adventures of Sherlock

Holmes. Whenever he was able, Edmond Locard interviewed professionals in different fields, even traveling abroad to find them. Born in 1877, he was utterly fascinated with criminal identification, believing that the future of criminal investigation lay in fingerprinting. However, while he recognized the value of ridge patterns on the fingertips, he knew they could be smudged or fragmentary—and even faked with sticky tree sap—so he examined the number and shape of the numerous pores that lay along the ridges, believing they might be more useful. There were between nine and eighteen per millimeter, and their patterns were as individual as the ridge patterns. Locard was convinced that only a few millimeters of a print was required to prove identity, and he called his new science poroscopy. But his initial fame came from another type of analysis, for which he received attention in 1912.

Locard had made numerous failed attempts to interest the police in Lyon with establishing a crime laboratory, so he devised a private lab of his own, in honor of his hero, Sherlock Holmes. In 1910, he acquired rooms in the attic of the Palais de Justice and then sought opportunities. In preparation to perform "police science," he purchased a microscope, reference books, and measuring devices, and studied various forensic techniques, especially the analysis of trace evidence. He believed that in any criminal encounter, offenders left traces of their presence, as well as carried some evidence away on their person. It might be so slight as to be nearly invisible, but Locard believed that a thorough inspection with the right tools could produce it. He once said "To write the history of identification is to write the history of criminology."

Few people listened to him until he took on a case the following year. Counterfeit coins were being used around town to buy goods, and the police thought they knew the identities of the culprits, but they had a difficult time nabbing them. Finally, three suspects were arrested and brought in. Locard got wind of the arrest and asked if he could examine their clothing. No one understood why he wanted to proceed in this manner, but since they had nothing else, they gave him permission.

Using a pair of tweezers, he went over one man's clothing to remove specks of dust from around the pants pocket area. He then turned his attention to the shirt-sleeves, brushing dust off them onto clean sheets of white paper. He then examined these samples with a microscope. Under magnification, it was evident that the dust contained characteristics that revealed its origin. Specifically, Locard looked for minute traces of metal and found them. Chemical tests applied to the dust grains affirmed what he observed, and the proportions of tin, antimony, and lead from the dust matched those in the counterfeit coins.

It only remained to find the same traces of metal on the clothing of the other two suspects, which he promptly did. That kind of evidence pressured the men to confess. Locard gained valuable publicity for his scientific identification of suspects. Within a decade, he would publish a book about the many cases on which he was consulted,

Policiers de roman et policiers de laboratoire (*Police in Novels and Police in the Laboratory*).

As Locard gained a reputation for detecting what the eye could not see, he was invited into more challenging cases. In one, Émile Gourbin, a bank clerk, was a suspect in the strangulation murder of his girlfriend, Marie Latelle. However, he had an alibi. At the estimated time of the murder, midnight, he'd been with friends, playing cards, and his friends all affirmed this. The location of the game was miles from the crime scene and they had been occupied until 1:00 A.M., at which time Gourbin went to bed. Either the time of death was in error or he was innocent. The police were stumped; they were certain he was the killer but could not budge his story. They needed undeniable physical proof.

When Locard examined the victim, he saw impressions deep into her neck, so he went to Gourbin in the holding cell and scraped under his fingernails. Under the microscope, he found in the debris tiny flakes of skin and noted that they had a pink tint. He tested them and found ingredients common to cosmetics. Looking to form a link with the victim, he examined her face powder under the microscope and found the same type of makeup as he'd removed from under Gourbin's fingernails. When he learned that the makeup had been custom-made for the victim, he proved that the substance was therefore unique, indicating that Gourbin had scratched the victim's skin.

The evidence impressed even Gourbin. He finally admitted that he had advanced the clock at his friends' house in order to dupe them and have an alibi. He had gone out to meet Marie after the others were in bed and when she refused to marry him, he had strangled her in a rage. The prosecutor insisted that he had clearly intended her death and the jury agreed.

Once Locard proved himself again and again, he received more staff and more funding. With greater resources, he went on to prove the value of dust and fibers for revealing what people have brushed against in their daily worlds, and in his manual he prescribed standard methods for handling such fragile evidence. It had to be done with care, and with tweezers and clean paper. Dust or dirt packed on to shoes should be removed carefully, layer by layer, with distinct substances kept separate from one another on sterile paper.

Another case gave him the opportunity to utilize his ideas about poroscopy. In June 1912 in Lyon, someone entered an apartment and removed four hundred francs and several valuable jewels. Fingerprints were left on a rosewood box that had contained the jewels. The Vucetich method from Argentina was applied and one of the prints was matched to a man named Boudet, who had a criminal record for burglary and theft. Another print matched his partner, Simonin. Confronted with the evidence, they nevertheless refused to confess.

Locard saw his chance, so he made enlargements of the prints to reveal the pattern of pores along the ridges. During the judicial sessions, he showed these photographs to the jury and explained how individual pore patterns were. He had found over nine hundred pores on the fingerprints left on the box and that number matched Boudet. The jury was impressed with Locard's meticulous work and found both men guilty, sentencing them to hard labor.

When Locard published his *Traité de Criminalistique*, he claimed to have solved seven cases that year by examining the fingerprint pores. He tells the story of a woman going to market with female underclothing who was knocked out with ether while riding on the train. When she revived she claimed that two men had stolen her merchandise and all her money. They had left the bottle of ether behind, from which the police managed to lift fingerprints. But they found that the fingerprints matched the victim and there were no other prints on the bottle. When confronted the woman admitted that she had attempted to commit suicide but had wanted to make it appear to be murder to save her family the shame. Yet while Locard continued to support poroscopy, it failed to catch on anywhere else. Nevertheless, his voluminous work on forensic investigation left a significant legacy in its own right.

ESTABLISHING SCIENCE

Even as the young scientist won fame in France, in 1911 the first fingerprint evidence was admitted into an American court. Charles Crispi, a.k.a., Caesar Cella, a burglary suspect, was arrested and sent to trial based on a match between his file and a fingerprint found on the window of a garment shop that had been burglarized. Detective Sergeant Joseph A. Faurot, a member of the New York Police Department since 1896, had long bought into the fingerprint theory and had hoped for a case to prove its worth in court. He'd been privileged to study this method at Scotland Yard and he sought to encourage its adoption across Manhattan. He testified that Crispi's print matched the one at the scene, so the judge asked him to demonstrate. He had jury members touch a window and a glass tabletop, asking one person to touch both, and then told Faurot to identify who it was. In less than five minutes, Faurot made the correct identification. Before he could go further with his proof, however, the burglar confessed, so the judge gave him a reduced sentence. Yet even without a dramatic resolution in court, Faurot's testimony provided fertile material for the press.

That was the year in which Alphonse Bertillon experienced his final humiliation before he died three years later. On August 21, a thief stole Leonardo da Vinci's famous *Mona Lisa* from the Louvre. He left fingerprints behind on the discarded case, and while Bertillon did keep track of fingerprints by now, his card system was still based on body measurements. Over the years, he had collected tens of thousands of cards in his system, which meant that with only a fingerprint as evidence, he and his staff had to search through each and every card, which took weeks. After the thief was finally arrested, thanks to a tip, and identified as Vicenzo Perrugia, it turned out that his card was in Bertillon's file as a repeat offender. It had simply been impossible to locate amid all the others. Thus it was that Bertillon (and everyone else) learned that his system, while innovative for its time, was too unwieldy to be of use. His successor lost no time in replacing it with fingerprinting.

At the time, most police officers did not rely on microscopes, but in 1912 that trend shifted. The microscope was the first scientific device to be used in a U.S. murder case, when a Massachusetts-based homicide was solved with the analysis of threads on a coat. Millionaire George Marsh, seventy-eight, had been shot to death on April 11, 1911, and dumped along an embankment in Lynn, Massachusetts. Yet the killer had not touched his wallet or gold watch, indicating the possibility of a personal motive. However, no one could identify an enemy.

At the crime scene, a few yards from the body, an investigator had picked up a swatch of material with a pearl-gray overcoat button fastened to it. Detectives located a landlady who offered information about Willis Dow, whom she had seen studying Marsh's house with binoculars. Another landlady turned over a coat that he'd left behind, and detectives saw that he'd removed all the buttons, so they had no way of knowing if the one they had found came from his—which would place Dow at the scene of the murder. They sent the coat to a textile school, where Professor Edward Baker microscopically compared fibers from a hole on the coat with the torn piece of a material from the scene. The colors and raw ends appeared to match, and the fabric swatch was the right texture and material to be from the coat. It was likely that when Marsh had pulled one button off, the killer had cut the others to prevent an identification.

Then a Colt .32 pistol was found in a canal near the crime scene and traced to William Dorr in California. It was a long shot, but determined detectives traveled to the West Coast and learned that Dorr was romantically involved with the victim's adopted daughter—and heir—there in California. That seemed suspicious, especially when Dorr turned out to be "Dow." They alerted the woman, who might well have been his means for acquiring Marsh's money, which meant she was in potential danger. With the threads from the coat, witnesses, the gun, and the circumstantial evidence, Dorr was convicted of the murder and executed.

Also in 1911, the Illinois Supreme Court had to confront fingerprint evidence when a case was appealed. The previous year, an intruder had shot and killed Clarence Hiller in his home. There were no eyewitnesses, but Thomas Jennings was apprehended near the Hiller home carrying a revolver. Since his clothing was bloodstained, he was arrested and soon the cartridges from his revolver were matched to unused cartridges found in Hiller's residence. In addition, four fingerprints had been left on fresh paint on a fence near the window through which the intruder had entered. Four experts testified that Jennings had left his own fingerprints on the fence. Around the same time as the New York–based trial featuring Faurot's testimony, the jury accepted the expert opinion and convicted Jennings of the crime, sentencing him to hang.

Jennings's attorney appealed the decision on the basis of the admissibility of fingerprint evidence. The appellate court now had a chance to examine this emerging issue. The justices arrived at a momentous decision.

Fingerprints had been admitted in Great Britain, the court said, and the relevant experts, who had written texts on the subject, had concluded that the evidence was reliable. They proved to have extensive experience reading fingerprints, so the court made a historic ruling that there was a scientific basis for the system of fingerprint identification, and that it was in general and common use by law enforcement. Thus, the trial court had been justified in admitting it. "Expert evidence is admissible," the court stated, "when the witnesses offered as experts have peculiar knowledge or experience not common to the world, which renders their opinions founded on such knowledge of experience, an aid to the court or jury in determining questions at issue." In this precedent-setting case, fingerprint evidence was deemed to have a scientific basis, and the notion of science gained an early version of its legal codification. Thus, the sentence stood and Jennings was hanged.

Four years later, New York's Court of Appeals faced a similar task with expertise in document examination in *People* v. *Risley*, a forgery case involving an attorney who had changed a record with a typewriter at his office. A mathematics professor offered a probability analysis on whether two words could have been typed on any other typewriter except the one in question. One appellate court allowed the testimony but the higher court reversed the conviction, rejecting the credentials of the witness and his method of analysis. The problem was that the professor was not an expert on the relevant issue: typing. In addition, he had offered only speculative opinion, not opinion based on observed data, and the laws of probability were not conclusive. One judge stated that courts should be "cautious" about scientific proof, and "cannot sanction it until its accuracy has been fully tested and vindicated by experience."

Over the next two decades, science would come to the forefront of legal proceedings, demanding that judges spell out just when testimony and evidence would be considered "scientific."

EIGHT

STANDARD PROCEDURE

IN LOCARD'S WAKE

Professor Victor Balthazard, a medical examiner, professor, and expert on trace evidence, pointed out in a 1913 article how the markings on fired bullets make them unique, because every gun had imperfections and peculiarities that ended up leaving distinct marks on the bullets. He described a case of homicide in which the victim had been shot several times. The police had a suspect but could not confirm that his gun was the weapon used. Balthazard took photographs of the murder bullets and bullets fired from the suspect weapon, enlarged them, and found eighty-five similarities. The suspect was convicted and Balthazard confirmed for medicolegal societies the instinct of physicians before him who'd performed similar experiments, his pronouncement arriving just as the first North American forensic laboratory based on

Locard's model was established in Montreal, Quebec, with *L'Institute de Medicine Legale et de Police Scientifique*. Around the same time, New York City replaced its British-based coroner system with a medical examiner, requiring death investigators to acquire training in medicine and pathology. Another invention that assisted law enforcement during this decade was Albert Schneider's vacuum apparatus for the collection of minute particles at a crime scene.

During World War I, Dr. Leone Lattes understood the significance of Landsteiner's blood group discovery and had shown in Turin, Italy, that these distinctions could be made even on bloodstains that had dried and aged. He used distilled water to make the blood liquid again and successfully applied Landsteiner's tests, so he set out to write *The Individuality of Blood*, in which he indicated how forensic investigation could benefit.

After the war, Luke May devised a dual-lens microscope that he named a "Revelarescope." With it, he entered the case of a child abduction in Roy, Washington, and was able to match a knife to cuts on pine needles. He also demonstrated the value of striation patterns for comparisons between specific tools, such as screwdrivers or chisels, and the marks they made. Proclaimed as an American Sherlock Holmes, May worked as a "consulting private detective" out of his Revelare International Secret Service, having participated in an investigation when he was only sixteen. Believing in impartial scientific investigation, he specialized in fingerprint, document, and firearms analysis, and he offered his laboratory services to the police. Later he would write for *True Detective Mysteries*, hoping to educate others in the importance of science. He would also be considered as a participant when America's first private crime lab did open, but that was still more than a decade in the future.

For forensic purposes, a police anatomist prepared the base for a bust in 1916 from a skeleton recovered in Brooklyn. He placed the skull on some rolled newspaper, placed fake eyes into the sockets, and covered the bone with plastic. A sculptor added sufficient details and this resulted in the identification of a missing woman. U.S. physiologist Thomas Hunt Morgan also demonstrated at this time that chromosomes carry inherited information, which would one day have implications for crime investigation.

EXPERTS AND GUNS

In the field of firearms and ballistics, a 1917 incident that initially seemed to have mostly local significance would figure into a more sensational case during that era. On a cold winter day, March 22, 1915, Charles Stielow, a German immigrant, came across the bodies of his employer, Charles Phelps, and Phelps's housekeeper, Margaret Walcott. Both had been shot to death with a .22-caliber revolver, and things had been taken from the house that indicated that whoever had killed them had probably eliminated them as eyewitnesses to a robbery. Stielow notified the police, and was then forced to testify in his broken English at an inquest. He claimed that he did not own a revolver. However, a private detective dug around and located a .22 that Stielow once had owned, which made him a primary suspect: Not only had he had access to the right type of weapon to have committed these murders, but he'd also lied. Detectives interrogated him until he finally confessed, believing if he did so he'd be allowed to go home to his wife, but instead he was held for trial. However, he refused to sign a statement, and once in jail, he recanted the confession. Nevertheless, he was tried for murder.

For expertise, the prosecution hired Dr. Albert Hamilton, who claimed that among his many professed specialties was knowledge of firearms. He testified that test bullets fired from Stielow's revolver showed an abnormal scratch in the barrel of the weapon, which he then matched to the bullets used in the double homicide. Hamilton even took photographs to further impress the jury, although he had to admit the scratches weren't visible in the photos. He got around that one by indicating that he'd accidentally photographed the wrong side of the bullets, adding that it took a certain amount of expertise to recognize them. When asked to point out the area on the gun barrel that was uniquely scratched, he said it could not be done because the bullet's momentum had expanded the bullet in such a way as to fill in the scratch with lead, rendering the scratches invisible. As nonsensical as it sounds in retrospect, the jury believed it and convicted Stielow, sentencing him to die.

The prison warden who observed him believed that

he was innocent, so he persuaded people with money to hire an experienced team to launch an investigation on the prisoner's behalf. They managed to pin the murders on two drifters, who confessed. An attorney, George Bond, looked further into this case, assisted by Charles E. Waite, an employee of the New York State prosecutor. Waite asked firearms expert Inspector Joseph Faurot, the same man who had brought fingerprinting to New York, to test the weapon. Right away, it was clear from the amount of dirt and rust on it that no one had fired it in years: not Stielow, and not Hamilton. Annoyed, they went ahead with the testing.

To prevent damage to the bullets, Faurot fired the suspect gun into water. Then optics expert Max Poser, from the Bausch and Lomb Optical Company, examined them under a microscope and could not see the alleged scratches that Hamilton had "observed." He then found that the bullets used in the murders had been fired from a weapon with an abnormal land-and-groove pattern—a manufacturer's defect in which the lands were quite wide—whereas Stielow's gun was normal, so it could not have been the murder weapon. It was clear from all of this information that, had a genuine expert examined the weapon in the beginning, that person would have excluded Stielow right away.

By this point, the unfortunate man had served three years in prison. He received a full pardon, but it did not escape Waite's notice that Hamilton's glib pseudoscience had nearly sent an innocent man to his death. It had also cost the county so much money that they declined to prosecute the real killers, who thus got away with a double homicide.

To prevent such incidents in the future and keep frauds like Hamilton out of the courtroom, Waite devoted himself to making firearms examination into a much more exact science. To achieve this, he set out to develop a base of information by cataloging all weapons ever manufactured since the first Colt revolver, in terms of construction, date of manufacture, caliber, number, twist and proportion of the lands and grooves, and type of ammunition used. He visited one gun manufacturing firm after another around the United States, including Colt and Smith & Wesson, and secured their assistance. It was a monumental task, perhaps more than he'd realized when he first made the decision, but after three years he had data on almost all of the types of guns manufactured since the 1850s. He saw that no type of gun was exactly identical to any other, and he was soon able to tell from which type of gun a bullet had been fired. He'd made a significant achievement, but then he realized that many guns available to criminals had not been made in the States at all but in Europe. Some two-thirds of total available guns were not included in his inventory.

Waite realized that he could either give up his ambition as unfinished or travel to Europe and continue to catalog all the guns he could find. To his credit, he decided to persist. At the end of his year in Europe, he had collected hundreds more models of firearms to test and catalog.

To look at the imprints left during the manufacturing process, Waite needed a good microscope, so he challenged Max Poser to develop a device for studying and comparing bullets. Poser came up with a microscope with fitted bullet holders and measuring scales. John H. Fischer, a physicist, then invented the lighted helixometer probe to inspect with a magnifier inside a gun barrel. The final instrument came from chemist Philip O. Gravelle, whose extensive work in microphotography inspired the comparison microscope, which combined two microscopes into a single unit for side-by-side comparison of two bullets under the same lens. Working together, they founded the Bureau of Forensic Ballistics in New York. A couple of years later, former colonel and army medical officer Calvin Goddard joined the team, but his true moment in history was still a few years away.

CONTRIBUTIONS

Back in Lyon, Locard had just published *L'enquete criminelle et les methods scientifique,* in which he stated that the act of criminality was so intense, it was impossible to act out without leaving a trace, or, as it's been rephrased, "Every contact leaves a trace." He was now increasingly concerned about faulty eyewitness identification. In 1920, a woman claimed to be Anastasia, a survivor of the attempt by Bolsheviks to slaughter the entire Russian royal family, the Romanovs. From news reports, people knew that Anastasia had been taken with her parents, her brother, and her three sisters to a building in Ekaterinburg, where they were shot multiple times. Once dead, the bodies were thrown into a mineshaft and an attempt was made to obliterate their identities with acid. However, after the remains were found, it appeared that two bodies were missing—including that of Anastasia. In Berlin, a girl saved from suicide indicated that she was the missing grand duchess. Many people who had known Anastasia were convinced when they met this girl that she had miraculously escaped the slaughter, but others had doubts. In Locard's *Proofs of Identity*, later published in 1932, he sided with the doubters. The physical features of this girl were too different from those of Anastasia, he said. In the end, he was proven correct, but not until a DNA analysis long after his death.

By now, many forensic cases were getting significant media attention. Since the mid-1800s, newspaper moguls had learned how well blood, scandal, and sex sold papers, and reporters sought out such cases. As sensational crimes garnered media coverage, it brought publicity to the innovations in forensic science and to specific figures who seemed to stand out from the pack of law enforcement officers.

One such case occurred in Camden, New Jersey, in 1920. Sixty-year-old David Paul, a bank messenger, had the task of delivering a satchel of money from his employer across the bridge to another company in Philadelphia, but along the way he disappeared. No leads were developed and more than a week later, hunters found Paul's body near a stream, buried in a shallow grave. An autopsy showed that he'd been beaten and shot. But there were some odd features to this murder: Although the ground was dry, Paul's clothing was wet, and it appeared that while he'd been missing for nine days, he'd been killed between twenty-four and forty-eight hours before he was found. However, aside from tread impressions made by car tires found near the victim, there were no other items of physical evidence to gather.

Investigators hypothesized that Paul may have decided to take the money and run, but had been killed in the process. Perhaps he'd had accomplices who then killed him. A few detectives thought that perhaps he had been kidnapped and held some place against his will before he was ultimately murdered. In any event, the satchel of money had disappeared and Paul's whereabouts for nine days, along with the solution to solving his murder, remained a mystery.

But the chief detective on this Burlington County case, Ellis Parker, was a persistent and often clever man, albeit rumpled and somewhat odd. He was convinced that the medical examiner's estimation was flawed. He returned to the crime scene to ponder the puzzle and came up with a better theory. He traced a pair of glasses from the scene to Paul's neighbor, Frank James, who appeared to be in cahoots with another man, Raymond Schuck. They'd been seen together spending large sums of money. However, both had alibis for the estimated time of Paul's murder.

At a dead end but not yet ready to concede defeat, Parker then looked at tanning factories along the river, and the water yielded a high content of tannic acid, a strong preservative. If the body had been submerged in it, he surmised, decomposition would have been significantly retarded, foiling the medical examiner's typical methods for determining time of death. Parker used these facts to confront James, who broke down and confessed that he and Schuck *had* dumped the body in the water but later retrieved and buried it. Thanks to Parker's persistence, both were convicted and sentenced to death. Parker became yet another man dubbed as the American Sherlock Holmes, in part due to his shrewd ideas and fascinating series of cases. He'd know more than a decade of accolades before he ended up undermining his career in a prominent case still to come.

BACK TO BULLETS

Another incident in 1920 highlighted ballistic analysis once again. On the afternoon of Friday, April 15, 1920, in South Braintree, Massachusetts, two men approached a couple of security guards who were delivering the payroll money for the Slater and Morrill Shoe Factory, and opened fire. According to witnesses, one of the men shot both guards mortally, and the other pumped several more bullets into them. They then took the payroll boxes containing nearly $16,000 and sped off in a black Buick. People nearby said they appeared to be Italian, and one man reportedly sported a handlebar mustache.

Investigators on the scene recovered six ejected shell casings from the sidewalk around the dead men and were able to trace them to three manufacturers: Remington,

Winchester, and Peters. They also found the apparent getaway car, abandoned, and they linked it to an earlier robbery. The mastermind appeared to be an Italian thug named Mike Boda, but when they located his hideout, he was gone, supposedly on his way to Italy. Yet two of his associates, Italian laborers who were part of an anarchist organization, fit the general descriptions of the robbers: Nicola Sacco, twenty-nine, and Bartolomeo Vanzetti, thirty-two. Sacco even had a handlebar mustache. They denied owning guns, but a search turned up illegal pistols on their persons. Sacco's was the right caliber for the murder weapon—a Colt .32 automatic. He also had two dozen bullets on him made by the three manufacturers whose bullets had been matched to the shells. Both men were promptly arrested. Vanzetti was found guilty of the robbery that had occurred before the double homicide, and Sacco was later tried with him for the murder of Alessandro Berardelli, one of the shoe company's security guards.

The trial began on May 31, 1921, and public opinion was clearly against the defendants, because they were perceived as dangerous men. Yet many foreigners, resentful of American xenophobia, sided with them and the Sacco-Vanzetti Defense Committee called the ordeal a witch hunt, with these men serving as scapegoats for America's fear of international politics.

Four bullets removed from the murdered payroll guards were delivered to the self-educated ballistics experts for both sides, and their task was to determine whether Sacco's .32 pistol was indeed the murder weapon.

The prosecutor's experts could not agree, while the defense experts, James Burns and Augustus Gill, exuded scientifically unwarranted confidence in their opinions.

Still, on June 14, Sacco and Vanzetti were convicted of murder and sentenced to be executed. The verdict was likely based on the fact that had little to do with what the experts or multiple witnesses had said: the bullet that had killed Berardelli was so outdated that the only bullets similar to it that anyone could locate to make comparisons were those found during the investigation in Sacco's pockets. The jury had even used a magnifying glass to examine the bullets for themselves and had finally bought the prosecution's case.

Right away another expert declared the others to be frauds, and his opinion was sufficient in 1923 to get a hearing for considering a retrial. To bolster their side, the defense team hired Albert Hamilton, who was adamant that the gun in the possession of the two men was not the murder weapon. Hamilton was the same man who had testified so ignominiously in the Charles Stielow trial, and shockingly enough, was still allowed into a courtroom.

For the prosecution, Charles Van Amburgh had reexamined the evidence, using new technology that enlarged photographs for better viewing. He offered photos of the fatal bullets and those known to have been fired from Sacco's revolver, finding them to be identical. However, he was unprepared for the cunning Hamilton.

The self-styled expert brought in Sacco's .32 and two new Colt revolvers. There in court, he disassembled them all and then tried to use one of the new barrels to replace the one from Sacco's gun. Judge Webster Thayer saw what he was doing and demanded he return the original barrel for Sacco's gun. Thayer then denied the motion for another trial. Hamilton had blown it, both for the defense and for himself.

Eventually, a committee was appointed to review the case, due to the persistent controversy and accusations of a miscarriage of justice. They contacted Calvin Goddard at the Bureau of Forensic Ballistics. In the presence of one of the defense experts, Augustus Gill, he fired a bullet from Sacco's gun into a wad of cotton and then put the ejected casing on the comparison microscope next to casings found at the scene. He looked at them carefully. The first two casings proved to be no match, but the third one was. Even Gill agreed that these two bullets had been fired from the same gun. The second original defense expert (not Hamilton but Burns) also concurred, and that clinched the case. That same year, Sacco and Vanzetti went to the electric chair. Vanzetti still claimed he was innocent, while Sacco declared, "Long live anarchy!" (Subsequent investigations with better technology in 1961 and 1983 both supported Goddard's findings.)

EMPHASIS ON EVIDENCE

While all of this was going on in Massachusetts, events of note occurred in other places. In 1921, Attorney General Harry Daugherty transferred the fingerprint files from Leavenworth prison to a location in Washington, D.C., which became a central fingerprint database for the entire country. In 1924, Congress established a national depository of fingerprint records at the FBI, which took custody of over 800,000 fingerprint files from various prisons.

In Australia, also in 1921, Charles Taylor, a food analyst working for the government, had shown a flare for solving mysteries with microscopic analysis. Early one morning, the body of a nude female adolescent turned up in an alley, the victim of a sexual assault. It appeared that she had been killed elsewhere and then placed in the alley, and oddly, she had been thoroughly bathed. A man from the neighborhood indicated that the owner of the wine shop, Colin Ross, had been pacing along the street the night before. When Ross was questioned, he said he had seen the girl and was able to describe everything she had been wearing. He knew her as Alma Tirtschke. Other people recalled that she had been inside the wine shop.

Detectives asked Ross to turn over two blankets, which were sent for analysis. Charles Taylor went over them with a magnifying lens, and on one he found twenty-one strands of reddish hair the same color as Tirtschke's; several were long enough to have come only from a female. He established with a microscope that the hair was human—in contrast to animal hair—and that it was similar in consistency and structure to the dead girl's hair but unlike that of other redheads that the suspect said he'd been with. In addition to a witness who claimed that Ross had confessed the crime to her, Ross did not have a chance. The case was a first in Australia for microscopic evidence in a criminal investigation and the first conviction based largely on forensic evidence.

Back in America, another fingerprint case grabbed headlines. It was a brisk Saturday morning, September 16, 1922. Around 10 A.M., two teenagers turned onto De Russey's Lane in New Brunswick, New Jersey, and spotted two people lying dead next to a crab apple tree. The police arrived to find that the male victim had been shot once in the head and the female three times. The feet of both were pointing toward the crab apple tree, and the woman's head rested on the man's right arm, as if posed. Her left hand rested on the man's right knee, and a brown silk scarf, soaked in blood and covered in maggots, wrapped around her throat. The man's right hand extended partly under the dead woman's shoulder and neck, and their clothes were perfectly in order. A .32-caliber cartridge case lay near the bodies, as well as a two-foot piece of iron pipe. Scattered pieces of torn paper, which turned out to be letters and cards, lay between them, along with a small card leaning against the heel of the man's left shoe. A man's wallet identified the male victim as Edward Wheeler Hall, forty-one, pastor of a local Episcopal church.

Shortly, Albert J. Cardinal of the New Brunswick *Daily Home News* arrived and picked up the card at the foot of the male corpse. It was Hall's business card. Spectators stripped the crab apple tree of its bark for souvenirs and the card was passed from hand to hand. No one thought to handle it carefully or preserve it as evidence.

The estimated time of death for both victims was

some thirty-six hours earlier. In the morgue, it was determined that the woman's throat had been cut from ear to ear. When people who knew Hall were questioned, it was clear that Hall had been involved in a rather public affair for the past four years with Eleanor R. Mills, thirty-four, a choir singer and wife to James Mills, a school janitor. The reverend was married to Frances Stevens, a wealthy woman seven years his senior. The obvious chief suspects were the spouses of both.

Frances Stevens claimed that she did not know of the affair, although on the night of the murder, Hall had said he was going to visit Mills to see about a medical bill. He was gone two hours when Frances's fifty-year-old brother Willie, who had a mental deficit that prevented him from living on his own, came out of his room to say goodnight. Frances then went to bed. At two-thirty A.M., she woke and went to the church in search of her missing husband, accompanied by Willie. The church was dark. In the morning, Frances called the police anonymously and learned that no casualties had been reported. She heard about the double homicide from a reporter and suggested that robbery was the motive, since Hall's gold watch was missing.

James Mills claimed ignorance of the affair as well and also believed robbery had been the motive. He said that his wife had gone out that night and at ten-thirty, he had gone to the church to look for her. She was not there, so he went home and went to bed. The next morning, without reporting his wife as missing, he went to work. At 8:30 A.M., he went to the church and encountered Mrs.

Hall, who mentioned that her husband had not come home the night before. He claimed that he asked her whether she thought that they had eloped and she replied, "God knows. I think they are dead and can't come home."

Two other suspects were the brothers of Mrs. Hall: William Stevens, known as Willie, was impulsive, explosive, and somewhat reckless. He admitted to owning a .32-caliber revolver, which he had not shot in over ten years. The older brother, Henry Stevens, fifty-two, was a retired exhibition marksman. He lived fifty miles away and claimed to have been out fishing when the murders took place.

Soon after, two bloodstained handkerchiefs were turned in to the police. One had no identifying marks, but the other was a woman's handkerchief, initialed in one corner with the letter *S*. Henry Stevens admitted it was his. Another discovery was a package of love letters from Hall to Eleanor, and Hall's diary. Mills sold them for $500 to the New York *American*.

Then Jane Gibson, the "Pig Woman," came forward to say what she had seen. She lived in a converted barn near De Russey's Lane and raised hogs. She told police that her dogs were barking around nine o'clock that Thursday night and she had seen a man in her cornfield, so she mounted her mule and went after him. She spotted four figures near a crab apple tree—two men and two women. Then she heard a sharp report and one of the figures fell to the ground. A woman screamed, "Don't! Don't! Don't!"

Gibson turned her mule away, heard a volley of shots, and saw another person slump to the ground. Then she heard a woman shout, "Henry!" She had also seen an open touring car parked on Easton Avenue, close to the crime scene, and the Halls owned such a car. A car coming into the lane behind her illuminated the people and she saw a woman in a long gray coat, and a man with a dark mustache and bushy hair walking together toward the abandoned farm. A little later, she said she heard a woman ask, "How do you explain these notes?" She also said she had seen a woman run away after the man was shot and the other two caught her and dragged her back, shooting her three times. Some time later, Jane went back to fetch a moccasin she had lost, and saw a big woman with white hair weeping over one body. Yet several people contradicted her, claimed she was a liar, and generally discredited her.

Four years went by with no breaks in the case until July 3, 1926, when Arthur S. Riehl, who had married Louise Geist, the maid who had worked for the Hall family, filed for annulment. He discovered that she had withheld information: She had told Mrs. Hall on September 14 that Hall had plans to elope with Mrs. Mills. She went with Mrs. Hall and Willie Stevens that night, driven by the chauffeur, and received $5,000 for keeping quiet about what she knew. Louise insisted that her estranged husband's tale was a pack of lies.

On July 28, warrants went out for the arrest of Mrs. Hall. She hired Robert McCarter, a well-known trial lawyer, to represent her, and he teamed up with Clarence E. Case as assistant chief defense counsel. The state appointed state senator Alexander Simpson as prosecutor. He interviewed Jane Gibson and announced his intention to proceed. Shortly thereafter James Mills admitted that he had indeed known about the affair and had threatened divorce.

Arrest warrants were then issued for Willie Stevens and Henry Carpender, cousin to Mrs. Stevens, who lived close by. Simpson contended that Mrs. Hall had been caught up in a murder, but did not commit it. The grand jury indicted Mrs. Hall, her two brothers, and Henry Carpender, and they were arraigned. Each pleaded not guilty.

The trial for Frances Stevens and her two brothers for the murder of Eleanor Mills was scheduled for early November. Both bodies were exhumed and new autopsies performed, which turned up evidence that Eleanor Mills's tongue and larynx had been cut out. That got the media's attention and the trial in Somerville, New Jersey, attracted reporters from all over. The evidence included Willie's fingerprint on the calling card found at Hall's feet, Mrs. Hall's admitted anonymous call to the police to inquire about "casualities," a brown coat of hers that had been dyed black after the murders (she claimed to have worn a gray one), and the fact that one of her private detectives was said to have tried to bribe a key witness.

Many witnesses came and went, quite often discredited. Three fingerprint experts testified that the left index fingerprint of Willie Stevens was on the calling card found at the scene. The most impressive witness was Joseph Faurot from New York City's police department, the detective who had assisted so ably on the Stielow investigation. He offered transparencies of Willie's fingerprint to compare to the print on the card, claiming there was no doubt that Willie had touched the card.

But his testimony was interrupted by news of the failing state of the Pig Woman. Her physician said her blood pressure and rising temperature would make courtroom appearances detrimental to her health, so the trial was adjourned for a few days.

The main piece of physical evidence, the calling card with the print, came under much fire, because the card had been exposed to the elements for thirty-six hours, had been passed from hand to hand, and had not been carefully handled as evidence. The defense attorney called it a fraud, without explaining why Willie's print was on it.

Then the Pig Woman was carried into the courtroom on a stretcher, with her mother shouting from the front row, "She is a liar! Liar, liar, liar!" Nevertheless, Gibson told her story—yet a third version—claiming that Mrs. Hall, Willie Stevens, and Henry Stevens were there on De Russey's Lane that night. (She seemed to have forgotten that in her earlier statements, she had seen only two people out there with the victims.) She had seen Henry Stevens and another man wrestling with a gun when it went off. Then she told how Mrs. Hall's detective had warned her to keep her mouth shut.

When the defense came on, they presented enough witnesses to make Henry Stevens's alibi credible. Since

Gibson had heard the name *Henry*, and Stevens was a marksman and a relative, he had been railroaded. Also, Henry Carpender was known to relatives as Harry, so Mrs. Hall would not have called out "Henry" to address him. With Henry pretty much cleared, Willie was next. He surprised most of the audience by holding his own with the prosecutor.

Then it came out that the first time Jane Gibson had seen the defendants, she was not able to identify them. A farmer, George Sipel, claimed that Gibson had offered him money to say he had seen her that night on De Russey's Lane, as well as two men and two women by the cars parked there. Thus, this star witness lost her credibility.

Mrs. Hall was next on the stand. Simpson went after her for the statements she had made to James Mills that she believed the two missing spouses to be dead. She said it had seemed obvious when they did not return home by that time. Simpson also wondered why she had not mentioned her nocturnal trip to the church and to Mrs. Mills's house until after a night watchman had reported seeing her enter her house, but he could not turn this into a telling issue.

By the time all was said and done, 157 people had taken the witness stand in this record-breaking trial, and even the *New York Times* had devoted ninety front-page articles to it. Those who believed in fingerprint evidence were certain that Willie was guilty, and along with him his sister. Yet American juries were not yet that impressed with such evidence. The jury took three separate votes before they reached a verdict, acquitting all three defendants.

No one else was ever accused of these crimes, although other suspects have been considered. No murder weapon was ever found, and the alleged handkerchief evidence led nowhere.

SHERLOCK'S MANY FACES

The next "Sherlock Holmes" was a chemist who proved his worth on a kidnapping case. On the night of August 2, 1921, Father Patrick Heslin disappeared from his home in Colma, California, in the company of a stranger who had indicated an urgent need of the priest. Soon a letter sent from San Francisco demanded a ransom of $6,500 for Heslin's safe return. The correspondent indicated that the pastor had been beaten unconscious and was dying, and described an elaborate arrangement that involved releasing chemicals to kill the priest when a candle burned out. The kidnapper promised another letter, but it failed to arrive. Fearing the worst, local police contacted Dr. Edward O. Heinrich.

As a professor of criminology at the University of California at Berkeley, Heinrich was versed in many areas of forensic science and had consulted in a number of cases that seemed to have come to dead ends, solving them with painstaking criminalistic analysis. He had also worked with August Vollmer, a police chief who had encouraged Sergeant John Larson to devise a machine that would measure deception via raised heart rhythms and systolic blood pressure. They would put it to the test in this case.

Heinrich arrived in Colma to study the ransom letter. He believed, from the decorative style, that the writer was in some trade, such as a baker. The police were skeptical, but Heinrich stood by his opinion. Still, it didn't offer much in the way of leads, so a considerable reward was posted. A week passed and a lanky Texan named William Hightower entered the archbishop's office to say that he'd heard from an anonymous source that Father Heslin was dead and buried. He'd found the spot while digging for bootleg liquor, and he was interested in the reward. Since Hightower was a baker, he became an immediate suspect. He dug himself in deeper when he led detectives to the burial spot, even indicating which end of the grave was the foot, and they unearthed the corpse; Heslin had been beaten over the head and shot twice.

Larson brought his lie detector to use on Hightower, who agreed to be hooked up. As he answered questions, sticking to his original absurd story, it was clear that he was lying. Larson believed that Hightower had murdered the priest.

Hightower's home was searched, and detectives found a canvas tent imprinted with the word *Tuberculosis.* Also, a jackknife removed from Hightower's pocket had microscopic shreds of white cotton, like that of a cord wrapped around a tent peg and piece of wood found with the body, as well as grains of sand similar to sand from the burial site. The same sand was also in the seams of the tent.

Heinrich compared the handwriting on the tent to that in the ransom note and found them to have originated from the same source; they also matched other notes that Hightower had penned. The part of the note that was typed had been done on Hightower's typewriter. Heinrich surmised that Hightower had wrapped the body in the tent and used the dreaded disease label to keep anyone from looking inside. With this collection of evidence, Hightower was convicted of murder.

Then on October 12, 1923, the number 13 train was blown up inside a tunnel in Siskiyou, Oregon, and four men were killed. Daniel O'Connell, chief of the Southern Pacific police, arrived to investigate. He learned that the fatalities were the result of a botched robbery by three masked men. They had shot the engineer, fireman, and brakeman, while leaving the mail clerk to burn alive in the exploded car. Investigators believed there were no productive leads from a battery linked to a detonating device, a pair of shoes, and a pair of greasy denim overalls found at the scene, but they arrested a mechanic, holding him on the most fragile circumstantial evidence: He might have provided the battery and he wore greasy overalls.

Heinrich was called in for a consultation and he examined the detonating device and the overalls from the scene. From these, he deduced that the clothing had been worn by a left-handed lumberjack who worked in the Pacific Northwest, was white, between the ages of twenty-one and twenty-five, less than five-foot-ten and weighed around 165 pounds. He also had small feet and light brown hair.

Fingernail clippings indicated he was fastidious. This description did not match the suspect.

To explain how he had arrived at the profile, Heinrich pointed out that the overalls were stained with pitch from fir trees, and their size, with the shoes, had provided the suspect's height and approximate weight. Wood chips in the right pocket indicated the position he took when cutting a tree; the overalls buttoned on the left, so he was left-handed; and hair caught on the overall button indicated his race, age, and hair color. Buried deep inside the bib pocket, overlooked by everyone else, was a receipt for registered mail. A magnifying lens offered a number, which led to a letter that implicated three D'Autremont brothers from Eugene, missing since the incident, and one of whom was a left-handed lumberjack the right size and age. Evidence from their personal effects matched evidence on the overalls, and handwriting on a gun receipt also confirmed their association with the crime. After a long manhunt, one brother was convicted while the other two confessed.

Heinrich also proved that Charles Schwartz, a man suspected to have burned in a laboratory explosion in 1925, was not the deceased; instead, a look-alike had been murdered and substituted in a staged death. He used an x-ray to indicate that the victim had been bludgeoned to death, but a missing molar that matched dental records contradicted Heinrich and affirmed that the victim was Schwartz. When Heinrich sought photographs of Schwartz, his widow reported that someone had entered her home and stolen them all. But then she found one that had been left at a photo studio, which she delivered. Heinrich noted that the corpse's earlobe, still intact despite the fire, was the wrong shape to be Schwartz's, and upon closer inspection, he realized that the missing tooth that identified Schwartz had in fact been yanked out recently and rather violently. The eyes, too, had been gouged out rather than burned in the fire, and chemical tests indicated that the fingertips had been burned with acid. To Heinrich, this was further evidence of a clever con artist who had thought of all the angles.

It turned out that Schwartz was in fact a long-time confidence man, living off his wife's money and lying about practically everything. He had befriended a missionary who resembled him, then killed the man, set an explosion in his laboratory to burn the body, and fled. Investigators lured him out of hiding by hinting in the papers that his wife had received a large insurance payment, but when he sensed the police closing in, he shot himself.

One other California-based case for Heinrich introduced a new type of evidence. John McCarthy, foreman for the Vallejo Street Department, entered his home on December 19, 1925, where he was shot in the chest. Before he died, he said, "I fired Colwell." The police believed he was referring to Martin Colwell, fifty-nine, a local ruffian with a criminal history. McCarthy had dismissed him from a street labor gang and during his drunken binges he'd been overheard making threats against McCarthy. Colwell was arrested with a .38 revolver in his pocket, while a .38 bullet that had passed through McCarthy was recovered in his home. One chamber in Colwell's gun was empty and he had three more bullets on his person. A box of ammunition found in his home was missing four, but Colwell said he'd been drunk at the time of the incident.

The gun and bullets were sent to Heinrich. He test-fired one of Colwell's bullets, several from his ammunition box, and others from an unrelated batch that were similar in caliber. Examining the bullets under a microscope, along with the bullet that had killed McCarthy, Heinrich found convincing similarities that led him to believe that Colwell had fired the bullet that killed McCarthy. However, prosecutors were concerned that this would prove insufficient in court, so Heinrich produced pictures from a stereoscopic microscope that showed the tiny scratches from the gun on the bullets in a side-by-side comparison, as a single three-dimensional image. He experimented over and over until he was able to click his two cameras simultaneously to produce the image. No court officer had ever before seen such an image, where the photographs of two different bullets seemed to perfectly merge. Heinrich called this a bullet fingerprint, and it was clearly a precedent in an American court. Yet the defense had a strong witness as well and despite the impressive photos that proved the bullets had come from the same gun, the jury hung.

The case went back to trial in just over a month, and

the jury members asked to look into the microscope to see for themselves what Heinrich had observed. He arranged for a demonstration, allowing each member to look through the lenses. They then asked Heinrich to reshoot the photograph for them. He accepted the challenge and had a darkroom set up near the courtroom. He managed to replicate his feat and the jurors were finally convinced. They sent Colwell to prison for life. Heinrich's approach inspired refinement of the equipment so that future scientists could offer similar results.

By this time, August Vollmer had founded a crime lab at Berkeley and persuaded the University of California to offer courses in criminal justice. He became known as the "father of modern policing," encouraging innovations and requiring police officers to get college degrees.

LIES AND SCIENCE

Throughout human history, it's been a common assumption that liars can be unmasked by their own bodily changes during deception, such as agitation, a flushed face, and decreased salivation. To take advantage of this evidence and to make lie detection scientific, in 1917 psychologist William M. Marston invented a measuring device. He claimed to be able to detect verbal deception through an increase in systolic blood pressure. It caught the attention of the FBI and the Department of War, who thought the device might be useful for the interrogation of prisoners of war. But to Marston's chagrin, in

1923, his invention inspired a far-reaching court decision in *Frye* v. *United States* that not only became the benchmark for the admissibility of scientific evidence but also proved to be bad news for the polygraph's future.

The defendant, James T. Frye, was convicted of murder, and he appealed on the basis that the court had not allowed an examiner to testify about the results of the "systolic blood pressure deception test," which he had passed. The federal circuit court examined the case and upheld the conviction, articulating the "*Frye* Test." The court decided that "just when a scientific principle or discovery crosses the line between the experimental and demonstrable is difficult to define. Somewhere in this twilight zone, the evidential force of the principle must be recognized, and while courts will go a long way in admitting expert testimony deduced from a well-recognized scientific principle or discovery, the thing from which the deduction is made must be sufficiently established to have gained general acceptance in the particular field in which it belongs."

Novel evidence, such as this device, which had not yet gained recognition among the relevant professional community, was thus excluded in court until a pool of experts could prove its value. At that time, the polygraph was new and had not gone through objective testing. Experts had no ground for claiming its reliability or validity, even though the test was being utilized.

Marston would continue to work on his invention, later introducing it in December 1941 through the character of Wonder Woman (through whom he also hoped to persuade young women to see how femininity could also be strong). This character utilized both a lie detector device and a "truth" lasso.

Even as Marston's device was blocked from the courtroom, another recruit to the Berkeley police, Leonarde Keeler, also built a portable lie detector. His first case involved a request to exonerate a co-ed accused of stealing and lying about the crime. He then tested everyone else in the sorority house and identified the thief as the house mother. She confessed.

To the systolic blood pressure measurement, Keeler added a way to also measure the skin's resistance to electricity, because during deception it was believed that resistance levels drop. Keeler continued to refine his device, hoping for a case to bring it positive publicity.

GREATER STAKES

During World War I, John Edgar Hoover, a Department of Justice lawyer, had been appointed assistant director for the Bureau of Investigation. After a shake-up in that institution in 1924 based on a scandal, Hoover took over as acting director. At that time, there were over four hundred special agents manning nine field offices around the country. Hoover had a vision for an elite force of professionals, so under his leadership, the department became more disciplined and was eventually renamed the Federal Bureau of Investigation, or the FBI.

During the same year that Hoover rose to power,

forensic evidence in the form of questioned document examination was highlighted in a shocking case in Chicago. Nathan Leopold and Richard Loeb showed the nation that it was not just thugs or petty offenders who committed crimes. They were two rich kids, arrogant and vile, who'd decided to commit the perfect murder just to prove that they could.

Leopold was enamored of the idea espoused by German philosopher Friedrich Nietzsche that no one can dictate morality to a superior man. Nietzsche proposed the idea of the *Übermensch* who made and lived by his own rules. Leopold persuaded Loeb that as intelligent young men with clever minds, they were among these exceptional beings, but they had to prove it with an act that would affront common morality. They started with cheating friends out of their money in card games, shoplifting, and committing arson and burglary—acts that thrilled them. But they weren't thrilled about the lack of press coverage, so they decided they had to do something much more dramatic. They came up with a plan to murder someone. Since they were smart, they believed they would get away with it and then be able to gloat secretly as the police ran around trying to solve the crime. Just the thought of this scenario made them laugh. They should have taken a cue from Raskolnikov's fate in *Crime and Punishment*: They weren't the first to erroneously believe in their own superiority.

On May 21, they went to select their victim, trolling around an exclusive boys' school. Along came fourteen-year-old Bobby Franks, so they offered him a ride. He climbed in the car and within a block, one of them bludgeoned him with a chisel to knock the boy senseless, and then shoved a rag into his mouth to smother him to death. Afterward they drove some distance away so they could strip him and pour acid on his face and genitals to prevent people from identifying him. Finally they tossed the naked and mutilated body into a culvert where Leopold often went birding, and then returned home to compose a ransom note to Franks's parents for $10,000. This was part of the game.

But the body was discovered the following day, before the boys were ready. Nearby, investigators picked up a pair of discarded glasses. Since they had a set of unique hinges, the glasses were traced to a local optometrist, who turned over his records. It wasn't long before the police came to Leopold's door, but he claimed he'd lost the glasses while birding in that area. While this lie set up a block, it was not the end of the investigation, mostly because the boys badly wanted people to know what they had done. It was some trick, taking credit but not getting caught. Loeb, who had also been questioned because Leopold gave his name as an alibi witness, offered theories to friends and reporters about the crime, suggesting that Franks was a perfect victim.

The police looked into the backgrounds of both young men, deducing that whoever wrote the ransom note was educated, and eventually they found samples of Leopold's typing, which matched the note. They did not find the portable typewriter in his possession (the killers had been smart enough to discard it), but when they caught the two men in a lie, it wasn't long before detectives used this to trigger a confession, getting each to turn on the other. The press reported this kidnap/murder for a thrill as being unique in the annals of American crime.

A major news organization tried to entice the famous Austrian psychoanalyst, Sigmund Freud, to travel to Chicago to offer his opinion, but he showed no interest. The judge listened to celebrity defense attorney Clarence Darrow list the "mitigating" factors, which failed to move him, but he was disinclined to sentence such young men to die, so he responded to Darrow's argument against the death penalty and gave both culprits life in prison. It's likely they discovered there that they weren't so superior after all.

MEASUREMENTS AND COMPARISONS

Even in Egypt, practiced observation made a contribution. The Scottish pathologist Sir Sydney Smith, who recognized the careful approach of Joseph Bell to medical diagnosis, served as the medicolegal advisor to the Egyptian government. During the early 1920s, a body was found, shot in the head and lying in the desert sand. Smith went out to observe the investigation. There were no clear tracks, but Smith witnessed how Bedoin trackers with refined visual skills had been enlisted to assist. They saw tracks and were able to discern that a man who came there had worn sandals, knelt where a rifle cartridge was found, and walked up to the body. Then he'd removed his sandals and run away barefoot. At a certain point, he had come to a car, and now there were four sets of tracks of men wearing boots. These led to an encampment of six men. Many different men were marched across an area of sand to compare their footprints, and the Bedoins identified a man who turned out to own a gun that was matched to a discarded cartridge casing. He also had a motive: The dead man had been having a dishonorable affair with his sister.

Forensic medicine had long had to deal with the size and nature of bullet wounds, specifically to determine whether someone had been shot accidentally, homicidally, or by his own hand. Exit wounds had to be distinguished from entrance wounds, and distances had to be measured between a weapon and a wound. Unburned particles of powder that move through the gun barrel toward a person leave certain telltale effects on skin and clothing, as do gases that penetrate the wound. Pathologists had these formulas nearly worked out when the advent of smokeless powder during World War I forced changes in the analyses. Despite the attempt to approach wound analysis with scientific precision, and the provision of laboratories with comparison microscopes and other devices that offered new technology, certain cases brought personalities together that defied the ability to remain detached. One such case in 1926 was the shooting death of Bertha Merrett.

John Donald Merrett lived with his mother, Bertha, in Edinburgh, Scotland, and on the sly he was stealing from her accounts. On March 17, he reported to a woman who worked as a maid in his home, Rita Sutherland, that his mother had just shot herself. (By some accounts, Rita actually found Mrs. Merrett.) Bertha was lying on the floor of the library with a wound to her head, but she was still alive. Taken immediately to the hospital, she was treated. In the meantime, Donald went out with his girlfriend, checking only to see if his mother was dead. Rita thought it strange that Mrs. Merrett had just been writing a letter and would then have shot herself. Donald told her it was due to "money worries."

Bertha managed to survive for a few days, and during this time she affirmed to a friend that she had heard a loud noise near her head and her son had been standing close by while she was signing papers, yet no one in authority questioned her for details. She was a "suicide case," which was illegal, so whatever she said was discounted as the product of a deranged or disoriented mind. Once she died two weeks later, her testimony was lost. But the son's behavior afterward, and the fact that he had inherited his mother's estate, alerted officials to the possibility that Bertha Merrett had been murdered. While Dr. Henry Littlejohn, who performed the autopsy, found that the entry wound and trajectory path of the bullet were consistent with suicide, other physicians were not convinced. Still, no one had evidence of foul play, so the case appeared to be closed.

Littlejohn had seen five hundred cases of suicide by shooting, so he had the experience to make a call, but he thought the circumstances, along with reports he heard about Donald, were suspicious. In addition, he'd found no powder burns under the skin, which was unusual for a close shot. He decided to discuss the matter with his former student, Sydney Smith, who had published *Textbook in Forensic Medicine* the year before. In it, Smith had stated that ballistics was a branch of forensic medicine, not something to be left in the hands of gunsmiths. Smith had done numerous experiments that measured shooting range and had observed under different conditions how bullets enter skin. In this case, he advocated using a suspect weapon and its ammunition to make comparison shots under conditions as close as possible to the original crime. Each weapon, he said, produced a characteristic pattern.

Smith asked Littlejohn to get the Spanish pistol used in the Merrett murder, along with its cartridges, and urged him to make some comparison shots from various distances. He could then examine these in the context of the powder marks left on the skin. Littlejohn fired the pistol at white cards from several different, carefully measured, distances. He then washed the sheets of paper, in case the doctors attending Mrs. Merrett had done so with her, and found that with close-range fire some residue stubbornly remained. There had been none on the victim.

Another doctor, John Glaister, assisted, and he and Littlejohn repeated the experiments with skin taken from an amputated human leg. When the Spanish gun was shot at close range, no amount of wiping would remove the powder burns. It was clear evidence that the gun had not been held close to the victim's head, and thus the death should not have been ruled a suicide.

Yet the defense attorney had cunningly engaged the service of celebrity expert Dr. Bernard Spilsbury, who had consulted with Robert Churchill, a gunsmith and impressive international ballistics expert. Spilsbury launched into a long explanation in defense of the claim for suicide. Yet rather than use the suspect pistol to conduct their tests, they had used one of similar caliber, concluding that any residue left on the victim from her self-shooting could easily have been wiped off at the hospital. Churchill explained the awkward trajectory path behind the ear as the female tendency to look away from a gun. Spilsbury said that he'd conducted a second set of tests using the Spanish pistol, but he'd substituted different ammunition and had fired at only the white cards. Even so, he got much more blackening with the original gun, which suggested that Littlejohn's experiments using the right ammunition and original gun had more validity. Still, he didn't change his opinion and Spilsbury's presence alone was persuasive, as he'd become one of history's first true celebrity expert witnesses. That meant his involvement could potentially add more weight to a case than it deserved.

The jury found that the case against Merrett was not proven, so while he had to serve time for a forgery conviction, he was released within a year. Nevertheless, twenty-five years later, he confessed to the murder, vindicating Littlejohn's claim that comparison tests done with a suspect weapon were the best means for making such determinations. Not only that, he is likely the person who killed two other people before finally taking his own life.

Another British ballistics case in Essex, England, involved George Gutteridge, a police officer, who was shot four times on September 27, 1928, and left on the road. Then human blood was found on a stolen car abandoned in London, and a cartridge case was located inside. Frederick Guy Browne, an ex-convict who operated a garage, was arrested. The police searched him and found twelve Webley revolver cartridges and a Webley revolver, which was loaded with ammunition that matched the discarded cartridge from the abandoned car. Browne had an association with William Henry Kennedy, another ex-con, who was arrested in a different crime and who attested to being with Browne when Browne shot Gutteridge as he attempted to apprehend them in the stolen car.

The imprint on the breech shield of the cartridge was compared against the revolver, offering a good match, but investigators decided to firm up the evidence. They microscopically examined around thirteen thousand revolvers to ensure that no other revolver had identical breech marks, and they found none. In addition, the ammunition in the revolver was a rare type, matching the bullets removed from the body, and the black powder discharge had a characteristic pattern that was also found on the body. The jury that heard this evidence convicted both men and they were hanged.

The St. Valentine's Day Massacre on a snowy February 14, 1929, led to the opening of the first private scientific crime detection laboratory in America. Seven men from George "Bugs" Moran's gang were waiting that morning around 10:30 A.M. in a redbrick warehouse for the S-M-C Cartage Company on Chicago's North Side. Three men in police uniforms and two dressed as civilians arrived in a police car and went inside. Moran happened along just then, late for the meeting, but when he spotted the police, he fled. It had been a set-up, and he suspected rival gangster Al Capone.

Area residents heard the blast of machine guns for several minutes, and then in the silence a dog howled. The police drove off and a few brave souls ventured into the building to see what had happened. They found seven bullet-ridden corpses on the warehouse floor, all shot in the back.

The uniforms made people believe that the police had staged this outright massacre against the unarmed men, and the incident became a scandal. Since the shooters had left behind seventy cartridge casings, allowing the weapons to be identified as .45-caliber Thompson submachine guns, it seemed possible to try to find the right weapons. Calvin Goddard arrived from New York as an independent investigator and fired each of the eight machine guns owned by the Chicago police. He then compared the results to evidence collected at the scene. No casings matched. That result was proof that a group of men had impersonated police officers to commit the murders. But it was difficult to discover who it was, and ten months passed before the police solved the crime.

A raid on the home of one of Capone's hit men produced two machine guns, which were sent to Goddard. He test-fired them and proved with microscopic analysis they were the weapons used in the warehouse bloodbath. That sent at least one of the St. Valentine's Day killers to prison and proved that the incident was part of a gang war between Capone and Moran. Evidently the men had been lured there to meet a nonexistent truck full of hijacked whiskey.

Goddard's careful work inspired two businessmen who had been on the coroner's jury to persuade him to run a private crime lab associated with Northwestern University in Chicago. He accepted, bringing ballistics, fingerprinting, blood analysis, and trace evidence under one roof. Luke May was invited to be a participant as well. This lab became a prototype for others around the country. When the FBI set up its own Criminological Laboratory in 1932, the agents consulted with Goddard.

The FBI acquired a comparison microscope and helixometer to complement its national fingerprint file, and the agency was now moving into the modern age of crime investigation. That same year, Thomas Gonzales applied the first test for gunshot residue on the hands of someone suspected to have fired a murder weapon, while at the Berlin Technical University in Germany, Max Knoll and Ernst Ruska had already developed and utilized the transmission electron microscope, which relied on a focused beam of fast-moving electrons to "see through" a specimen. This invention enabled scientists to see much smaller objects than had been possible with a light microscope, enhancing the science of trace evidence examination. But the next headline-grabbing case involved many different types of evidence analysis, as well as a tragic death that affected people the world over.

NINE

MURDER AND MEDIA

ELUSIVE IDENTITIES

On a cold March evening in 1932, in a recently constructed luxury home near Hopewell, New Jersey, Charles and Anne Lindbergh were preparing for the night. Lindbergh had won fame with his solo flight across the Atlantic Ocean, and he was considered a representative of America's best. Thus, he was a vulnerable target as well, and on this night he would learn the price of fame. The Lindberghs typically returned on Mondays to Anne's parents' home in New Jersey, but they remained in Hopewell on this particular Tuesday because their baby, twenty-month-old Charles Lindbergh, Jr., had a cold. As Lindbergh read in the study, he heard a loud *crack*, like a box falling, but he attributed it to the high wind blowing outside.

Upstairs, the baby had just been placed into his crib in the second-floor nursery. His nurse returned to check on him and discovered the crib empty. She asked Lindbergh if he had the child, but he did not, so he checked with Anne and they ran to the nursery. The child was clearly gone and an envelope on the windowsill told the story: The "princeling" had been kidnapped.

During those years of the Great Depression following the stock market's collapse in 1929, kidnapping was fairly common. Over the past three years around the country, there had been more than 2,500 such incidents, terrorizing America's wealthy. Lindbergh was just one more victim.

Although several police departments got involved, Lindbergh ran the show. Given his financial means to hire experts, the investigation would become a showcase for several areas of technical expertise; it also showcased some inept handling.

On the night of the kidnapping, the Hopewell police searched the house and grounds, finding a carpenter's chisel near several foot impressions leading from a ladder that had been used to access the nursery window. Less than a hundred feet away, a wooden extension ladder lay on the ground in three sections, one of which was split along the grain. Farther away, near a small dirt road, police found tire tracks, but no one thought to measure their width or make a cast. The same neglect was exercised on a footprint located in the wet ground below the nursery window, although one officer compared his own size-nine shoe and found the print to be larger.

Inside the nursery, detectives opened the note to reveal a threatening message:

Dear Sir!

Have 50000$ redy with 25 000 $ in 20 $ bills 1.5000 $ in 10$ bills and 10000$ in 5 $ bills. After 2–4 days we will inform you were to deliver the Mony.

We warn you for making anyding public or for notify the Police the child is in gut care.

Indication for all letters are singnature and 3 holes.

At the bottom-right-hand corner was a drawing of two interlocking, penny-sized, blue circles. The areas where they intersected had been colored red, and three small holes were punched into the design, at the left, right, and center. That would prove to be the signature.

In trying to reconstruct the incident, investigators could not understand why the kidnappers had taken such a risk, rather than wait an hour or two until it was certain that everyone was asleep in the house. The fact that the dog had been removed that evening to another part of the house brought suspicion on the domestic servants; they were also aware that the Lindberghs had changed their plans to return to New Jersey at the last minute. Yet some investigators believed it could easily be the job of an outsider. The house had been featured in magazines all over the country, complete with floor plans, so it was easy enough to learn where the nursery was, and whoever had brought the chisel had not known that the window shutter could not be locked. In addition, the family's movements could be easily monitored from the woods.

On March 5, a second letter with the same interlocking signature scolded Lindbergh for involving the police, and upped the ransom demand to $70,000. In all, fourteen such notes would be received over the course of the investigation. They were signed with the same symbol and often contained the same handwriting, misspellings, and grammatical errors. One note agreed on a go-between, a retired teacher named John F. Condon, and another gave instructions for the type of box that must be made for delivering the money.

Condon met their representative, but before offering money he insisted on a token of proof, and soon the baby's sleeping suit arrived at the Lindbergh home, with the following note:

> *Dear Sir: Ouer man faill to collect the mony. There are no more confidential conference after we meeting from March 12. Those arrangemts to hazardous for us. We will note allow ouer man to confer in a way like befor. circumstance will note allow us to make transfer like you wish It is impossibly for us. wy shuld we move the baby and face danger. to take another person to the place is entirely out of question. It seems you are afraid if we are the rigth party and if the baby is allright. Well you have ouer signature. It is always the same as the first one specialy them 3 holes.*

On the reverse side was:

Now we will send you the sleeping suit from the baby besides it means 3$ extra expenses because we have to pay another one, please tel Mrs. Lindbergh note to worry the baby is well. we only have to give him more food as the diet says.

You are willing to pay the 70000 note 50000 $ without seeing the baby first or note. let us know about that in the New York-American. We can't do it other ways because we don't like to give up ouer safty plase or to move the baby. If you are willing to accept this deal, put these in paper.

I accept mony is redy.

Ouer program is:

After 8 houers we have the mony received we will notify you where to find the baby. If there is any trapp, you will be responsible what will follow.

In preparation for identifications, fingerprint expert Dr. Erastus Mead Hudson applied a silver nitrate solution to the baby's toys to get a set of the child's fingerprints. These were photographed with a special fingerprint camera. He then worked on the ladder and came up with over five hundred prints, many of which were full latents. Eight were clear enough to be useful, but aside from one that belonged to a trooper, none could be matched to anyone.

The IRS offered a brilliant idea: Put the ransom

money into gold certificates, because the government would be recalling them, and these could then be used to track the kidnappers when they tried to spend them. The prescribed box contained $50,000, and a second package contained the additional $20,000. The bills were not marked but the serial numbers had been secretly recorded, and the smaller package contained large bills that would be easy to spot.

On March 31, Condon handed over the money, but he received only another note to the effect that the baby was being held on a boat at Horse Neck Beach, near Elizabeth Island. Lindbergh got into his plane to search for this elusive boat, while the Treasury Department distributed the serial numbers of the ransom money to area banks. The boat was not located, so Lindbergh returned home, disheartened. He and his wife could only hope that someone was taking good care of their child.

Then on May 12, 1932, at 3:15 P.M., a truck driver, William Allen, stopped on Princeton-Hopewell Road about one-half mile outside Mount Rose. He walked into the woods about seventy feet and saw a small white skull and a leg sticking up out of the ground. He reported this and the remains were identified from their size and the style of the garments as Charles Lindbergh, Jr. The cause of death was a massive fracture of the skull. The child was dead and the killers had gotten away. Once this discovery made the news, it seemed likely the kidnappers would stop communication, and they did. But the investigation continued.

The ransom and instruction notes were sent to several analysts, who all concluded that they had been written by the same person. The misspellings and English mistakes were consistent, as was the odd inversion of letters like *g* and *h,* and there were references in some to earlier notes or events, so the collection had internal consistency. They also believed that the same instrument had been used to punch the holes that created the signature symbol, and stated that the letters had been written on the same kind of paper with the same ink. The writer's nationality, inferred from the phraseology, was probably German. One expert, Albert S. Osborn, the author of the definitive *Questioned Documents*, composed a paragraph for the police to use with suspects, to ask them to write out certain words that could be compared with the notes, such as *our*, *place*, and *money*. But Osborn insisted that the paragraph had to be dictated, not copied, or a comparison would be pointless.

From time to time, as ransom bills turned up, they were examined for trace evidence. Most had been folded in a certain manner, and New York's toxicology lab found particles of glycerine and emery, which pointed to the possibility that the individual in possession of the money used an emery wheel to grind tools. Many bills had lipstick or mascara marks, and a few had traces of blonde, red, and brunette hair. The bills also bore a musty odor, as if they had been stored, but none yielded discernible fingerprints.

Investigators hoped to use Leonarde Keeler's polygraph, for which he claimed an accuracy rate of 90 percent, to interview Lindbergh's servants. However, Lindbergh refused to allow it. It was not clear whether he did not wish to believe anyone in his employ might be guilty or whether he simply did not accept the polygraph's purported reliability.

But Keeler did have his shot just then with another case. As the Lindbergh investigation wore on with no new leads, a reporter invited Keeler to participate in another incident involving a murdered child. It was a cold case from 1911 that had occurred in Madison, Wisconsin. Seven-year-old Annie Lemberger went missing from her bedroom and three days later her body was found in a pond near Lake Monona. The police quickly arrested a neighbor, John "Dogskin" Johnson, who had a record of molesting children. He confessed to the crime but then retracted it, saying he'd felt pressured by the threat of mob violence. Still, he had offered details that the police had withheld from the press, such as the fact that the child's nightdress was torn. Then townspeople began to have doubts, wondering how Johnson could have opened a window and made off with a girl her size without anyone in the house hearing it. The press ran stories about this "real-life mystery."

New suspicions centered on Martin Lemberger, Annie's father, and rumors spread that he had killed her by hitting her. Nevertheless, there was no evidence and Johnson remained in prison for the next decade. Then Johnson's lawyer, Ole Stolen, got him a hearing for a potential pardon, because he had located a witness, May Sorensen, who was a neighbor to the Lembergers. She claimed that Annie's mother and brother had admitted that Martin had indeed struck the girl, accidentally killing her.

Since the statute of limitations for manslaughter had expired, Lemberger was not prosecuted, but Johnson gained his freedom. Then on November 16, 1932, Robert Bishop, from the *Chicago Daily Times,* managed to arrange for Keeler to give the lot of them polygraph examinations. Lemberger, Johnson, and Sorensen all agreed, and Johnson and Sorensen in particular hoped to find support from the "lie box," because they expected money from Johnson's lawsuit against Wisconsin for false imprisonment.

On January 6, 1933, in the Lorraine Hotel in Madison, each person was hooked up, one by one, to Keeler's machine. Keeler started with Annie's parents and found their responses to be truthful. Johnson did not fare as well, but the results were inconclusive, so Keeler decided that he might have "guilty knowledge" of something connected to the incident. But it was May Sorensen who proved to be lying. When confronted, she alleged that Johnson's defense attorney had paid her to help him win the case. In addition, an examination of Annie's autopsy indicated that the child had died from asphyxia, not a blow to the head. Lemberger was exonerated and Keeler won publicity and fame for his machine, but the questions raised about Annie's murder remained unanswered.

During this time, America and other parts of the world were struggling through economic collapse. In the United States, the Eighteenth Amendment was repealed, ending Prohibition but bringing little relief. A series of bank robbers with names like Baby Face Nelson, Machine Gun Kelly, and Pretty Boy Floyd came on the scene, glamorizing this crime and bringing entertainment in the form of newsreels and media stories. But soon Bonnie and Clyde were ambushed in Louisiana, shot more than fifty times each, and police gunned down the notorious John Dillinger outside the Biograph Theater in Chicago. An examination of his corpse revealed that he'd attempted to have his fingertips surgically removed—which only made them even more unique.

In Germany, Adolf Hitler declared the Third Reich, setting out to restore his country's power, and three years later Spain entered a civil war. Italy and Japan acquired leadership in Benito Mussolini and Hirohito, respectively, and these countries collectively shared resentment toward the World War I Allies. World tension increased, and the Lindbergh case, likewise, neared its explosive resolution.

BACK TO LINDBERGH

On September 15, 1934, a gas station manager in Manhattan called in the license plate for a man who had used a gold certificate to pay for gas. The car belonged to Bruno Richard Hauptmann from the Bronx and a search of his wallet turned up another gold certificate from the ransom money. Coincidentally, it was discovered that he had quit his job on the day the ransom had been paid. Yet

Hauptmann maintained that a man named Isador Fisch had given him the money before departing for Germany, where he had died not long afterward. Once Hauptmann had realized that Fisch was dead, he had used some of the money, because Fisch had owed it to him. No one believed him and he was arrested.

During interrogation, Hauptmann was instructed to provide samples of his handwriting, as well as to copy the ransom notes as closely as possible (the method that Osborn had explicitly said not to use). Detectives made him write for hours, until he fell asleep at the table. He wrote his statement seven times, and nine sheets of dictated writing were taken to Osborn's son, Albert D. Osborn, also a handwriting expert. Hauptmann had been required to write with three different pens, with some samples written upright and some slanted. The end result was that there were more discrepancies between some of his writing samples than between his samples and the ransom notes. Osborn examined the statements, initially unconvinced that Hauptmann was the writer of the ransom notes, but he kept samples for further analysis. The police then instructed Hauptmann to write more statements, inappropriately dictating the spelling of certain words. "I was told to write it exactly as it was dictated to me," Hauptmann would later claim in court, "and this included writing words spelled as I was told to spell them." (In Hauptmann's other writings, such as letters, he did not misspell these words.)

He was also told to copy photostats of the ransom notes, although this was improper protocol, and at no time was he offered a lawyer. That same day, the police unearthed a large cache of ransom money hidden in Hauptmann's garage. This information was inappropriately conveyed to the Osborns, and that same day they identified Hauptmann as the author of the ransom notes. According to the analysis, Hauptmann had a peculiar way of writing the letters *X* and *T*. He also wrote *not* as *note*, using an open *O* and an uncrossed *T*, and wrote *the* in a strangely illegible manner. The most telling evidence was a diagnosis of agraphia—a peculiar "tic" of adding unnecessary *E*s onto various words, evident in both the notes and in his own letters.

However, another handwriting expert who saw exemplars from Fisch said that Fisch had written the ransom notes. There was also a statement by a witness to the effect that Fisch was seen near the Lindbergh home a short time before the kidnapping, which corroborated Hauptmann's Fisch story. Hauptmann had learned about Keeler's polygraph and begged to be tested to prove his innocence, but since a federal appeals court had already banned the results of systolic blood pressure deception tests from the courtroom, his attorney thought it was of little use.

Hauptmann's trial opened on January 2, 1935 in Flemington, New Jersey, at the Hunterdon County Courthouse. The New Jersey Attorney General, David T. Wilentz, led the prosecution. The *New York Journal* paid for defense attorney Edward J. Reilly's services in exchange for rights to his story. Reilly was assisted by a Flemington attorney, named C. Lloyd Fisher, who was inexperienced in criminal defense. Reilly had spent only thirty-eight minutes with the defendant and had publicly stated that he believed Hauptmann to be guilty. He also consumed alcohol at lunch throughout the trial, and his afternoon performance was, by many accounts, notably lacking.

Both Osborns gave damaging testimony, but Hilda Z. Braunlich, a European handwriting expert who believed that the ransom notes had been overwritten with changes that amounted to forgery, was not even allowed to testify. She later claimed that Reilly actually ordered her to leave the state. Also, the police had cleared Hauptmann's home of his writings, so the defense team was unable to offer specimens that might indicate discrepancies between his personal writings and the ransom notes.

The most impressive area of forensic identification involved the kidnap ladder. Wilentz described how an upright from the ladder fit nicely into a hole left in the attic of Hauptmann's home where a plank of wood had been removed, and put several wood experts on the stand. Arthur Koehler, of the U.S. Forest Products Laboratory in Madison, Wisconsin, testified that a semi-skilled carpenter had constructed the ladder from four types of wood: pine from North Carolina was used for the uprights, Douglas fir from the Western United States for an odd piece on the left upright, birch for the dowels, and Ponderosa pine for the rungs. The odd piece appeared to be from floor boarding and bore four apparently recent nail holes. He showed that the planing machine used on it had a revolving set of eight knives that comprised the cutting head, and one of them had a nick that left a distinct mark. A side cutter had also made a distinct impression. Koehler obtained the addresses of nearly sixteen hundred timber mills from around the country and requested samples of planed pine from those with the right type of machinery. These he examined microscopically until he found the right type of flaws. When he identified the likely mill for the wood's origin, he learned that they had shipped lumber to a retailer in the Bronx. The mill also verified that the wood had been part of a batch that was cut and planed there.

The defense argued that the ladder had gone through many hands since its discovery and that not only was the beam's placement in the ladder upright questionable, the police officer who ran that part of the investigation could have removed it from Hauptmann's attic, since he had access after Hauptmann had moved out. (Officers who had examined the attic earlier had not noticed a missing beam.) Several defense witnesses also discredited Koehler: One used Hauptmann's wood plane to demonstrate that holding a plane at various angles makes different marks. They ridiculed Koehler's claim that only Hauptmann's plane could have made the marks on the ladder. Then a general contractor and a millworker/carpenter testified that one could not match rail sixteen to the attic via grains because many grains in North Carolina pine looked alike. One man produced two pieces of board that looked similar in grain but one had been in a building for forty-seven years and the other for five. He also pointed out that the knots in rail sixteen were different from the other attic boards. The nail holes, said to fit perfectly with the attic beams, seemed to have been freshly made, because there was debris in the holes. They did not appear to have had nails in them for years, as the other attic boards did. Also, Koehler had admitted that if this board had been pried from a floor, there would have been marks on it from the nail heads, but there were none. In addition, the plank, when placed in the attic's gap, was thicker than the attic floor by one-sixteenth of an inch, and Koehler had no answer for that. Clearly, the wood testimony was not as impressive as it had initially seemed.

However, these defense experts could not refute that Hauptmann had made a purchase at the lumber company to which Koehler had traced shipments from the mill. It was circumstantial, to be sure, but nevertheless a strong piece of the puzzle. Yet the prosecution faced other problems. Hauptmann's shoe size was between a nine and a nine and a half, the same size as the officer who had compared his own shoe against the footprint at the estate on the night of the kidnapping and found it larger than his own. Thus, it was only Hauptmann's print if he had worn larger shoes.

The principal defense witness was Hauptmann, whom Wilentz attacked in long cross-examination, and who presented himself as arrogant and defiant. In a high, whining voice (nothing like that described by Condon about the man he had met), he was unable to satisfactorily explain Isador Fisch, the handwriting similarities, or his whereabouts on the night of the kidnapping. He claimed he had worked as a carpenter at the Majestic Arms Apartments until 5:00 P.M. on the day in question, but the prosecution produced work records that contradicted him.

After 29 court sessions, 162 witnesses, and 381 exhibits, the case was given to the jury at 11:21 A.M., Wednesday, February 13, 1935. Eleven and a half hours later, the jurors returned a unanimous vote of guilty. Hauptmann was sentenced to be executed. But this case, forensically speaking, was far from over.

Partly at the urging of Ellis Parker, the detective who had gained fame from the tannic acid case in Camden, Governor Harold G. Hoffman took up Hauptmann's cause. (Oddly, Parker and his son kidnapped a disreputable lawyer, Paul Wendel, and forced him to confess that he had kidnapped and murdered the Lindbergh baby. Wilentz questioned Wendel, found that he had been coerced by Parker and his partners, and dismissed the entire "confession" as a travesty. Parker and his associates were sentenced to prison for kidnapping.)

Hoffman met with Hauptmann in his cell and gave him a temporary stay of execution. He believed that Hauptmann should be given a polygraph, so Anna Hauptmann traveled to Chicago to talk with Leonarde Keeler, who was eager to be involved. To demonstrate the test's reliability, he used Anna as a subject. Reading off a list of numbers, Keeler correctly judged her age from the machine's recording of her physiological response. She was impressed, and returned to New Jersey to tell the governor that Keeler would provide his services at no charge and in secret. However, Keeler could not resist leaking the news, since he was desperately hoping for a noteworthy case, which violated the terms of the deal. He thereby undermined his chance to make a potentially monumental contribution. Instead, William Marston, whose machine had already been banned from the courtroom, was brought in to perform the test. He said it would take two weeks to get results, and arrangements were about to be made when the trial judge abruptly denied Marston access to Hauptmann. Time had run out, so Hauptmann was duly executed on April 3, 1936, by Robert G. Elliott, who had operated the electric chair for Sacco and Vanzetti nine years before.

THREADS, BONES, AND BUGS

Science was definitely having its day in court, even in the far reaches of Scotland. Helen Priestly was only eight when she set out to purchase bread for her family in Aberdeen, Scotland, in April 1934. She showed up at the bakery but failed to return home, so area residents who knew the family set out to look for her. The only immediate neighbors who declined to assist were Alexander and Jeannie Donald. Helen's body was finally located, stuffed inside a cinder sack on the ground floor of the tenement building where she had lived. Yet she clearly had not been in that spot a half hour earlier, so someone had placed her there. In fact, lividity indicated that she had died earlier and had been moved. Although she was clothed, her underwear was missing and there was blood between her legs. An autopsy revealed that she had been asphyxiated and sexually molested with extreme brutality, so every male in the vicinity was questioned.

A roofer said he'd heard a girl scream around 2:00 on the previous afternoon, which corresponded to the various signals of her time of death. When pathologists failed to find semen in Helen's vagina, they were no longer certain that the girl had been raped. In fact, they wondered if the "rape" had been staged by a female in the building, since the males all proved to have alibis for the time period when Helen was thought to have been murdered.

The person found to have lied about her whereabouts that afternoon was Jeannie Donald, who had been feuding with Helen's family and disliked the little girl. The Donald apartment was searched and the police turned up an apparent bloodstain, although tests revealed that it was a less sinister substance. Jeannie's husband had a solid alibi, but Jeannie did not. The Aberdeen police invited Professor Sydney Smith from Edinburgh University into the investigation, renowned for his work with ballistics. He had advised Littlejohn on the Merrett shooting.

Smith looked at everything and decided to focus on the sack in which Helen had been placed. He noted similar sacks, made in Canada, in the Donald apartment, and inside this one, he found plenty of trace material to examine: dust, fibers, and hair, both human and animal. Helen had not shed the human hairs, since they had been processed with a permanent chemical, but they proved a match to Jeannie Donald's hair. Smith also subjected the fibers from the bag to spectroscopic analysis, and he was able to match two dozen different fibers to materials in the Donald residence. He could not find the same comparison results in other nearby residences. In addition, bloodstains on a washcloth and several other items were the same type as Helen's, but did not match Jeannie.

Given the nature of Helen's wounds, Smith believed that there would have been bacterial contamination. He involved a university colleague, Dr. Thomas Mackie, who performed several analyses and found the same type of bacterial stain inside the dead girl as on a bloodstained washcloth in the Donald home. This provided both circumstantial and physical evidence against Jeannie Donald. In fact, her trial proved to be highly technical as the scientists explained their tests and their findings. Smith offered more than two hundred exhibits, which impressed the jury sufficiently to return a quick guilty verdict. Since it seemed clear that Jeannie Donald had probably reacted to the child's typical taunts rather than planning the murder, she received life in prison instead of death.

In Edinburgh, Scotland, on September 29, 1935, a young woman leaning over the railing of a bridge on the Gardenholme Linn spotted a cloth-wrapped bundle snagged against a rock. As she focused on it, she noticed the shape of a human arm protruding from a tear, so she ran for the police. A search soon turned up more human parts: two heads, parts of legs and arms, and one armless torso that proved at least one set of remains were those of a woman.

These parts had been wrapped in four separate packages, and some items from the other body were still missing, notably the torso, but investigators judged that these remains were female as well. A piece of newspaper from the *Sunday Graphic*, wrapped around two upper arms, showed the date of September 15, two weeks earlier. According to some sources, however, the police initially estimated the time of death as September 19.

Professor John Glaister from the University of Glasgow and Professor James Brash from the University of Edinburgh, invited to examine the remains, noted that the person who had killed and dismembered these two women knew something about bodies and how best to take them apart with a knife. That indicated either a butcher or someone involved in the medical profession. He had even skinned the faces, removed the eyes from one and the teeth from both to make identification difficult, if not impossible. From a pair of hands recovered, the fingertips had been removed. Glaister sent these seventy pieces to the University of Edinburgh's Department of Anatomy, treating them against further decomposition with ether and formalin. The maggots that infested these pieces went to Alexander Mearns at the University of Glasgow.

The first task was to separate the pieces and lay them out as individual corpses. A few more parcels were found, providing more parts for the grisly puzzle. The forensic team managed to determine that one woman had been taller than the other by half a foot, and the shorter one was still missing the torso, although the hands that seemed to go with her offered fingerprint possibilities.

Using the skulls, the team, which included Drs. W. G. Millar and F. W. Martin, determined that the ages of the victims were about thirty-five to forty-five for the taller one (stabbed five times, beaten, and strangled) and under thirty for the shorter one (possibly stabbed or bludgeoned). The undeveloped wisdom teeth and cartilage development for the latter finally set her age closer to twenty. Glaister surmised that the tall woman had been the intended victim, possibly killed during a violent rage by someone who knew her, and the other had probably been in the wrong place at the wrong time, seeing or knowing something that made her death inevitable.

Professor Mearns devised a timetable that established how long the maggots had taken to reach their stage of development. He combined weather factors, known insect behavior, and information about larval stages to estimate the postmortem interval at twelve to thirteen days, which indicated the bodies had been thrown into the ravine around September 16.

Detectives decided from the newspaper used to wrap the dismembered parts that the women had been killed in either Morecambe or Lancaster. They then learned that in Lancaster, a medical doctor, Buck Ruxton, had reported his thirty-four-year-old wife missing. The police were already acquainted with the household, including the fact that they had a young nursemaid for the children, because Isabella Ruxton had told them that her husband beat her and persistently accused her of infidelity. It soon turned out that Ruxton had also informed the parents of his nursemaid, Mary Rogerson, age twenty, that she had gotten pregnant so his wife had taken her away to get an abortion. He urged them not to go to the police but they did so, anyway, because Ruxton's tale failed to ring true to them.

Ruxton was now under intense suspicion, and more clues soon fell into place. When the Rogersons were shown a patched blouse that had wrapped victim parts, they recognized it as their daughter's. Then a charwoman who worked for the Ruxtons reported that Ruxton had told her not to come to work on Sunday, September 15. The next day she had found the home in a state of disarray, with carpets removed, unusual stains in the bath, and a pile of burned material in the yard. Neighbors said that they had helped to clean the house and had received a bloodstained suit and some carpets. They said Ruxton had told them that he'd cut them himself. However, traces of human fat found in the bathtub drain belied this, and on October 13, he was charged with the murders.

The police believed, from reports of Ruxton's behavior the week prior to the killings, that he had decided that his wife had cheated on him and that Mary had assisted in the cover-up. The case seemed entirely obvious, but Ruxton's attorney, Norman Birkett, insisted that since the bodies had not been definitively identified, his client could not be tried for homicide. Rogerson's fingerprints identified her, and the remains, based on anthropological calculations from the limbs, matched her height. The task remained to identify Isabella Ruxton, and for that they used a photograph of her and a unique new technique. Professor Brash photographed the skull of the unidentified body from the same angle as the photograph he had acquired of Isabella. He made this image into a transparency and superimposed it over Isabella's photo, showing how they matched on all the key points. The same was done on Mary Rogerson. Using a gelatin glycerin mix, Brash also made models of the left feet of both victims and proved that they fit the shoes of the missing women.

With all of the witness reports and physical evidence, and with the bodies now identified with this unique process, Ruxton was tried for both murders and convicted in just over an hour. On May 6, 1936, he was hanged. Glaister then wrote a book, *Medicolegal Aspects of the Ruxton Case,* which was published the following year. Sadly, the trunk of Rogerson's body was never located.

OTHER ANALYSES

Besides examining handwriting or a document composed on a typewriter, the analysis of questioned documents also involved an analysis of the type of paper used, and the state of this science proved useful in a 1934 New York case. Twelve-year-old Grace Budd had disappeared six years earlier in the company of a middle-aged man named Albert Fish. He had ingratiated himself with Grace's parents as "Frank Howard" and convinced them to allow him to take their daughter to a birthday party for children. That was the last they saw of either of them. But years later they received a letter from Fish, who fiendishly described how he had killed Grace, dismembered her, and cooked her into a stew that he had savored for days.

The Budd family turned this letter over to the police and a dogged investigator, Will King, traced the monogrammed stationery envelope to a former tenant of a flophouse on East Fifty-Second Street. The tenant identified his stationery and told them where he'd left it, in a closet. His handwriting did not match that from the note, so King looked into the possibility that someone else had used the abandoned stationery. But the next person who had inhabited the room had already moved on. The police asked the landlord if this person ever came around and he said, yes, he returned to claim checks sent there from one of his sons. The handwriting from the letter proved to match samples saved earlier from "Frank Howard," so detectives staked out the flophouse until they caught Fish on December 12.

Under arrest, he confessed in lurid detail, admitting to the crime and adding others. He'd also killed before, in 1910 in Delaware. Believing himself to be Christ, and obsessed with sin and atonement, he had made a practice of beating himself with spiked paddles. He also stuck needles into his groin, threaded rose stems into his urethra, and lighted alcohol-soaked cotton balls inside his anus. Fish admitted to the murders of three other children, but also claimed as many as four hundred others. Despite his obvious psychopathology, the jury convicted him of murder and sentenced him to death. Had an observant detective not paid attention to the clues provided by the stationery, Fish might have gotten away to commit yet more unspeakable crimes.

The same can be said for a case that involved fiber analysis. In 1936, Nancy Titterton, a novelist and the wife of an NBC executive, Lewis Titterton, was murdered in their Manhattan brownstone. Two furniture movers found her body. She had been strangled with her pajama top and left in the bathroom. All indications were that she had known her killer, and when there appeared to be few clues except some twine used to bind her and some green paint on the bed cover, Dr. Alexander Gettler from the toxicology lab was brought in. He examined the bedspread in meticulous detail and found one strand, half an inch long, of stiff white hair, which he soon identified with a microscope as horse hair. He said that it had to have been transported to the bed on someone's clothing, since there was no source for it in the room.

Since two furniture movers had delivered a horsehair couch that morning and it was those men who found the body, the detective in charge, Chief Inspector John Lyons, speculated that one of them might have paid an earlier call. In fact, it turned out that the furniture movers had been at the home the day before, picking up the couch for upholstery repair. Lyons identified the likely culprit, John Fiorenza, who had a criminal record, and then found a connection via the piece of twine: It had sufficiently distinctive markings to be traced to a manufacturer and distributor, which sold that type of twine to the upholstery shop where Fiorenza worked. Using this collection of evidence to pressure the suspect in a game of psychological cat-and-mouse, Lyons finally got the confession he needed for conviction.

The years 1937 and 1938 saw more improvements in forensic methods. Walter Specht did experiments with the molecule luminol, first synthesized in 1853, and saw that it offered a luminescent reaction in the presence of blood. If someone tried to wipe clean such evidence, an investigator could use luminol to locate the position and size of the former stain. Then scientists at the University of Kharkov developed a simple thin-layer chromatography, which enhanced toxicological analysis. With this process, a sample was subjected to a liquid solvent that separated it into its constituent parts, making it possible to directly analyze without resorting to the tedious process of extraction.

The next year, W. M. Krogman published his *Guide to the Identification of Human Skeletal Material*, ushering in the modern era in forensic anthropology. In fact, by this time, Russian paleontologist Mikhail Gerasimov had taken charge of the department of archaeology at the Irkutsk Museum, although he was barely out of his teens. He wanted to perfect the art of facial reconstruction, so he experimented with fossilized skulls in his care, calculating formulas for the thickness of musculature formerly on the face. His methods assisted others to learn facial sculpture from a skull, rescuing the field from the humiliation it had suffered in 1913 when two anthropologists proposing their own techniques had developed strikingly different faces from the same skull. While Gerasimov's emphasis on facial muscles contrasted with the American approach of calculating skin tissue thickness, both had developed more reliable procedures.

In 1940, there were also several notable innovations, according to several forensic timelines. Vincent Hnizda, a chemist with Ethyl Corporation, is believed to have pioneered the forensic analysis of ignitable fluids, assisting better arson diagnosis, while Hugh C. MacDonald, chief of the civilian division of the Los Angeles Police Department, devised the Identikit system, inspired by Bertillon's *portrait parlé*. MacDonald had been tracking down criminals in Europe who were taking advantage of wartime conditions, and he found that victimized people were offering only vague descriptions of offenders. He tried capturing these descriptions with facial sketches, but as cases piled up he sped the process along by creating standard transparencies of different facial features, which he could then lay over a facial outline. This worked well enough, so when MacDonald returned to the States, he consulted with colleagues to improve his collection and then approached a company that could mass-produce the images.

The earliest kits, which MacDonald claimed could compile 62 billion composite facial images, used more than five hundred "foils," or transparencies, on which separate features such as a nose, mustache, mouth, or set of eyes were drawn in varying shapes and sizes (initially, there were no ears). These pre-drawn images could be stacked on top of one another as witnesses described or picked them out, until the composite image satisfied the witness. Police departments could also quickly alert those in other jurisdictions by simply telegraphing the coded card numbers for each foil. It was a vast improvement over the older methods.

In addition, Bell Labs used L. G. Kersta's work on voiceprint methodology to produce the sound spectrograph. It analyzed sound waves from people's voices and produced onto a graph a visual record of patterns based on frequency, intensity, and time. The idea that someone could be identified by the sound of his voice had its origins in the work of Alexander Melville Bell (father to Alexander Graham Bell). More than one hundred years earlier, he had developed a visual representation of the spoken word, based on pronunciation, and he had insisted that there were subtle differences among different people. While Kersta's machine would eventually prove useful in crime investigations, during World War II it appeared to be more valuable for evaluating enemy communications.

As the war ended with two devastating atomic bombs dropped on Japanese cities, the world entered the atomic age. Science was now a valued tool for aggression and defense, and this elite status filtered into forensic science as well. It had come a long way from the early days of legal resistance. But in some ways, the stakes had been raised.

TEN

INTEGRATED INVESTIGATING

DEAD GIVEAWAYS

Someone entered the Chicago apartment of Josephine Ross in June 1945, hit her over the head, and cut her throat, but before exiting, the intruder placed tape over the wounds. The crime was soon linked to another such intrusion in October, wherein Frances Brown was murdered, mutilated, and then bathed in the tub. On a wall, written with the victim's lipstick, were the words *"For heaven's sake catch me before I kill more. I cannot control myself."* Not long afterward in January 1946, a six-year-old girl disappeared from her bed and a ransom note was left in the room that soon proved pointless: Her head and body parts were found in the sewer. The police increased their vigilance, but there were no more similar murders.

As reporters kept the story alive, they interviewed handwriting analysts. Just four years before, Albert S.

Osborn, who had written *The Problem of Proof* since his 1910 volume on questioned documents, had founded the American Society of Questioned Document Examiners. The issue at stake in this case was whether the ransom note had been penned by the same person who had left the lipstick message on the wall. Some experts said yes, and some said no. They would have to wait and see.

In June 1946, an off-duty cop spotted a young man surreptitiously entering an apartment, so he chased him down, unmasking him as William Heirens, a seventeen-year-old university student with a long record of burglaries and arsons. Under truth serum and a painful spinal tap, Heirens confessed to the three murders, although in later years he said that he'd been hypnotized and coerced.

Now the handwriting analysts were asked more formally for their opinions. George W. Schwartz compared the lipstick message and ransom note to handwriting in Heirens's school papers, declaring that there were no significant similarities. Prosecutors then sought another opinion, hiring Herbert J. Walter, who had participated in the Lindbergh case. He had told reporters that he did not see similarities between the ransom note and lipstick message, but after he examined the samples up close, along with Heirens's papers, he stated that Heirens was indeed the "Lipstick Killer." Walter believed that apparent discrepancies merely showed that Heirens had attempted to disguise his handwriting (seemingly a common judgment in this discipline). As Heirens blamed "George Murman," an alter ego, he pled guilty, receiving three life sentences.

A couple of bite mark cases in Britain put forensic odontology on the map, in a better light than handwriting analysis. Professor Cedric Keith Simpson, a pathologist for the home office who would gain the reputation of performing more autopsies than anyone in the world, was also a pioneer in forensic dentistry. He'd made his public debut in 1942 with the case of a beheaded murder victim found under a slab, her legs severed at the knees. From a list of missing persons, a potential victim was selected, and her family offered information about her doctor and dentist. Simpson utilized the dental plate of the intact upper mouth to track down a dentist who had done work on this middle-aged woman. The records matched, so she was identified. When her cause of death was ascertained as strangulation, her estranged husband was soon identified as her killer.

In 1946, Simpson also matched the distinctive teeth and weapon impressions from victims to a sadistic killer. On June 20, Margery Gardner was found dead in a London hotel, and she'd been whipped with a metal-tipped implement, as well as bitten, and raped with a blunt object. Neville Heath, twenty-nine, had signed for this room. The police found him in a seaside town, posing as a war hero, where he had met, dated, and killed a pretty young woman named Doreen Marshall. A braided whip among his effects matched to the bruise patterns on the first murdered woman. Found guilty, Heath was sentenced to be executed. But soon there was another bite-mark case that some scholars indicate was the true start of forensic odontology, at least in England.

On New Year's Day in 1948, a young woman turned up murdered, her body lying in a carpark outside a dance hall. Her name was Mrs. Gorringe, and some time during the course of her revelries the evening before, she had been bludgeoned and strangled. On her chest was a distinct bruise that turned out to be a bite mark, with two upper teeth and four lower teeth impressions visible. Her killer had left his calling card. Since witnesses had seen the dead woman arguing with her husband the night before, he became a suspect.

Simpson required Gorringe to submit to having his teeth imprints set in wax molds for making comparisons. The case closed when the man's teeth proved to have made the bite marks, and circumstances convinced the jury that he'd killed his wife.

But as a pathologist, Simpson participated in an even more notorious case the following year that did not involve dentistry, but did require a rather astute observer. It was the disappearance of Mrs. Olive Durand-Deacon in 1949 that put the police on the trail of an unusual killer in Crawley, England. She resided at the Onslow Court Hotel in South Kensington and had reportedly met a man, John George Haigh, who also lived there and who discussed a potential business deal with her. He had directed her to his warehouse, where he would show her the product he could make with her funding. But she failed to return to the hotel, and he himself went to the police with one of her friends. He readily admitted that he'd intended to meet Mrs. Durand-Deacon that afternoon, but she had failed to show up at the designated time. He had not seen her since and was concerned for her safety. Despite the suspicious circumstances, he seemed a decent enough sort. Yet that's not what some of the hotel staff had to say.

As detectives looked into the matter, they found that Haigh had a criminal record for swindling. When they searched his supposed warehouse, they found the receipt for a fur coat from a dry cleaner. Tracing it and asking more questions, they learned that the coat had belonged to the missing woman. Indeed, they soon found that Haigh had pawned some of Durand-Deacon's jewelry as well, right after she was last seen, so he was arrested on suspicion of murder. His first question concerned his chances of getting out of the local institution for the criminally insane, a signal that he intended to launch a mental illness defense so he could end up there.

He admitted to having killed six people compulsively since 1944. He said that he lured them into a storage area, bludgeoned them, and cut them to drink their blood. Then he would dissolve the corpses in a large acid-filled drum. He taunted the police with the fact that Durand-Deacon no longer existed. "How could you prove murder," he gloated, "when there is no body?"

However, Simpson was not so easily dissuaded. Aware that the dissolved remains of Mrs. Durand-Deacon had been poured out onto the ground at Haigh's supposed warehouse, Simpson believed he could find evidence. He took a team to the crime scene to search an area of sludge three inches deep. The task seemed nearly impossible, not to mention odious, but he managed to pluck out a gallstone and a partially dissolved foot, along with bone fragments, dentures, a lipstick container, and part of a purse. Under careful scrutiny, Simpson identified the bone fragments as human, and some showed an ailment that Durand-Deacon was known to have. The foot fit one of her shoes and her dentist recognized her specially made dentures.

In addition, behavioral evidence confirmed that Haigh had long been in the habit of entering fake business schemes wherein he would kill his dupes and acquire their property, which he then sold. Despite his reports of blood-drenched nightmares, it was clear to investigators and mental health professionals alike that each time he committed murder he had acted with rational awareness, selecting well-to-do victims, manipulating them, and killing them for his own enrichment. It did not take long for a jury to accept the physical evidence, reject the mental illness defense, convict Haigh, and sentence him to die.

SHARING THE WEALTH

Back in America, a group of professionals had been busy forming an organization for the collective forensic sciences. On January 18, 1948, two scientists had met at the police academy in St. Louis, Missouri, to discuss a project on which they had worked over the past year: the First

American Medicolegal Congress. Most professionals in the field were operating independently and these men believed that dialogue among toxicologists, biologists, chemists, criminalists, and other members of the scientific community who also worked in the legal arena would assist the field to become more interactive and holistic in its approach. This conference would mark the first international multidisciplinary effort to try to ground an organization. There had been two earlier conferences in St. Louis in 1945 and 1946 for coroners, attorneys, physicians, and investigators, but this current project was aimed at a far broader community.

Rutherford B. H. Gradwohl, director of research at the St. Louis Police Department and director of his own private medical lab, had originated the idea and offered financial support. Dr. Sidney Kaye, former assistant director and toxicologist at the St. Louis Police Research Laboratory, and now with the Office of the Chief Medical Examiner in Virginia, agreed to be the secretary-treasurer and co-planner. They had another partner as well, Orville Richardson, a St. Louis attorney, who served as their legal counsel.

For the inaugural meeting, they had invited scientists from other countries, but at the last minute, due to political disruptions and national financial problems, seven professionals from Cuba, Chile, Argentina, and Columbia were barred from leaving their countries. So on the evening of the conference at the Sheraton-Coronado, the schedule had to be revised, highlighting an acute loss: the co-chair, Dr. Israel Castellanos of the Cuban National

Bureau of Identification. He had participated in discussions over the years about the need for bringing together forensic scientists from all disciplines to resolve growing problems in the field. Alas, he could not be there to witness the results of his efforts.

The hope was that scientists and investigators from the different forensic disciplines would interact as colleagues, finding personnel for filling job openings, helping to resolve problems, and brainstorming solutions to unresolved crimes. In addition, they would support training for more personnel in areas that were sorely lacking, such as chemistry. The organizers sought to connect a community of professionals who could enhance one another's knowledge and careers.

One hundred fifty participants were present as the conference took place at Twelfth and Spruce Streets on January 19. For three days, twenty-nine papers were presented and Gradwohl articulated the purpose of the organization. Among the items discussed or described were the "intoximeter," the most recent blood analysis tests, improvements in microscopy, the reading of bloodstains, sexual criminals, and war crimes. Leonarde Keeler made a presentation on the polygraph, and toxicology provided plenty of topics on poisoning and drug overdoses. There were representatives for firearms and trace evidence analysis, and the renowned pathologist Milton Helpern commented on unexpected natural deaths. There was even a discussion about the relationship of trauma to cancer.

The group found this gathering to be helpful and agreed to meet annually, so a committee formed to seek the opinions of other scientists, attorneys, and jurists about the need for such a national organization. They were also to organize the program for the following year. With the adjournment, the mood was positive.

Later that year, a few members met to discuss an appropriate name for the fledgling organization. From among five titles, the final choice was not from the list, but the organization officially became the American Academy of Forensic Sciences (AAFS). Among their goals were to revise the standards of investigative techniques and the quality of testimony in court, engender public confidence in the judiciary, and create confidence in the courts in scientific evidence. In addition, they hoped to encourage the exchange of information among the various scientific fields, "encourage enlightened legislation," and improve the way scientists participated in attaining justice. As things progressed, the initial body became even more enthused about what they envisioned as an enduring association that would set and informally sanction ethical and procedural standards for their participation with law enforcement.

The second annual meeting occurred in January 1950 in Chicago's Sheraton, and the Windy City became the organization's home base. At this time, along with a constitution and bylaws, seven sections were established, including pathology, psychiatry, toxicology, immunology, jurisprudence, police science, and questioned documents. More would be added as the need arose, such as odontology, anthropology, and entomology, and a few section names would change. A committee drew up the first newsletter as well, published later that year.

Among the participants of this second conference was Mrs. Francis Glessner Lee, who had designed the "Nutshell Studies of Unexplained Death" during the 1940s. A philanthropist known to be domineering, exacting, and full of good humor, she had founded and supported Harvard University's weeklong seminars on homicide investigation, taught twice a year, and she posed her eighteen miniature scaled models of crime scenes (now on display at the Baltimore medical examiner's office) as vignettes of different types of violent death. On these, new officers could practice their theories. Mrs. Lee was a fan of Sherlock Holmes stories and she had created the models from a combination of fictional narratives, news stories, crime scene photos, and witness interviews. Her hope was to establish better contact between law enforcement and the medical community, so she fit right in with the AAFS agenda. To make her creations authentic, to the tiny pieces of furniture she collected internationally from dollhouse manufacturers (or commissioned from her full-time carpenter), she added sweaters and socks that she'd knitted on straight pins, tiny cigarettes she'd rolled, and even items she'd whittled. The settings were exquisitely detailed and would have served quite nicely as dollhouses, save for the blood and corpses inside.

Mrs. Lee included items that weren't clear at a glance, such as a tiny bullet stuck in a wall or a blood smear, but would help to train detectives and police officers in looking for subtle clues. What appears to be a suicide, for example, may change when a key item is noted—a baked cake, a load of freshly laundered clothing, an ice cube tray beside the body. Other scenarios included a bound prostitute with a sliced throat, a burned corpse, a man hanging in a wooden cabin, a woman in a bathtub with the water flow hitting her face, and a family murder-suicide. Detectives training on these models were forced to look carefully in the same way they would have to look at an actual crime scene. (Some of the models are still used for this purpose.)

The AAFS increased its membership during that decade from ninety to more than four hundred members. While there were few women in the forensic sciences, those who were involved became active participants in this organization, and its reputation for integrity, along with its collection of impressive professionals injected forward momentum. It would prove to be a major force in the establishment of forensic science.

BIOLOGICAL EVIDENCE

In late spring of 1948, four-year-old June Devaney was grabbed inside a children's ward of Queen's Park Hospital in Lancashire, England. Her body was found on the grounds within two hours, raped in a way that suggested irrepressible mania. The local police requested the help of New Scotland Yard, and among the clues were fibers on a windowsill, blood and hair from a wall, fibers on her body, and a pubic hair on her genitals. In addition, there were stocking prints in the ward, along with more fibers. During the investigation, a bottle was found that had been removed from its usual place, and on it were fingerprints.

The police began systematically to take fingerprints of local males for comparison to those found on the bottle. Within a month they had checked more than two thousand people associated in any manner with the institution, but they failed to turn up a good suspect. So they set out to get more ambitious and undertake something that had never before been done: They decided to fingerprint every male over the age of sixteen in the town. One by one, they went through hundreds and then thousands of men. On August 12, they found a match with the fingerprints of Peter Griffiths, twenty-two, who admitted after his arrest that he had committed the shocking rape/murder. Since he had a history of mental disorders, his sanity was questioned, but he was nevertheless convicted.

Similarly demanding, another British murder case involved a laboratory experiment for its solution. In 1949, a bundle found in the Essex marsh yielded a man's dismembered torso. By this time, forensic pathologists were often called to such scenes, and Dr. Francis E. Camps, from the London Hospital Medical School, responded. A painstaking albeit ambitious professional, he tended to dismiss coworkers as "scabs" and to compete with other pathologists, especially Keith Simpson. They squared off more than once in the courtroom, and Camps's egocentrism often undermined him more than anyone else, but he nevertheless solved some difficult cases.

After an examination, he concluded that the corpse in the bundle had been in the water approximately three weeks, placed there approximately two days after the stabbing death. For identification, Camps removed the skin from the hands to get a good fingerprint impression, and he soon learned that the deceased was forty-seven-year-old Stanley Setty, missing for two and a half weeks—directly after he'd cashed a large check in London. It was clear from the autopsy that one of his wounds had produced a large amount of blood, which indicated that there could be a significant bloodstain wherever he'd been killed. In addition, Setty's body showed postmortem injuries that suggested he'd fallen from a great height—such as from a cliff or a plane.

Investigators checked the small airports in the area to see if anyone had flown with large parcels. They learned that around that time a man named Donald Hume had hired a small plane and loaded two large packages onboard. Detectives from Scotland Yard entered this plane and identified bloodstained areas inside. In addition, they soon learned that Hume had known Setty.

The detectives found and questioned Hume, but he had a ready response: Two men had hired him to fly the parcels over water and drop them. As preposterous as the story sounded, the investigators knew they'd have to prove it was not so. Seeking another angle, they went to Hume's apartment. There, the housekeeper said that recently Hume had refinished all of the floors and worn out the cleaning implements.

At this point, Camps reentered the picture. He went

to Hume's apartment to look for areas of blood that would indicate that Setty had been stabbed to death in this place. Dr. Henry Smith Holden, a chemist and expert in serology, accompanied him. They already knew from tests that Setty's blood group was O, and despite finding only a few spatters Holden soon found human bloodstains of that type on three floorboards and in several rooms. Camps was not one to give up, so he kept looking and noticed cracks between the floorboards. He was certain he could see blood inside, so he and Holden ripped up the floor. They found what they needed, and when they had scraped together as much blood as they could, it amounted to about a cupful.

But they weren't finished. Now they had to make an important calculation. In order to compute how much blood had been spilled in order to leave the quantity of coagulated blood they'd found inside the cracks, they replicated Hume's floor in the lab and poured varying amounts of blood over this model. Keeping careful measure, they determined that three pints of blood had been spilled in that area alone, much more than someone having an accident or nosebleed would lose. Since adult human males have about ten pints of blood, a man losing three would be in bad shape, if not dead.

This result helped to prove Hume a liar. Had he merely brought the parcels containing Setty's remains to his apartment in preparation for the flight, and Setty had then bled out this much, Hume certainly would have noticed. But he hadn't said that the parcels had leaked in any way.

Nevertheless, the case proved to be a struggle in court. Camps and Holden had no difficulty showing that a bloody corpse had been in Hume's home, but they could not prove that he had committed the murder. Thus, he was sentenced as an accomplice, receiving much less prison time. In fact, when he got back on the street and knew the court could no longer touch him, he sold his confession to a tabloid, saying that he had indeed killed Setty, exactly as the scientists had surmised.

Another man who got away with murder for quite a few years—in fact, half a dozen murders—was finally caught with careful forensic analysis, and this much-reported case provided another opportunity for the scientists. The more they could prove themselves in cases that gained public attention, the better.

The landlord at 10 Rillington Place in London's Notting Hill had an empty flat in 1953 as the result of John Reginald Halliday Christie leaving, so he allowed the upstairs tenant, Beresford Brown, to use the kitchen. Brown noticed a bad smell, so he began to clean. It occurred to him that he might install a new shelf on the wall for his radio. He knocked on the walls to find the proper place and heard a hollow sound. Pulling away the wallpaper, he found the door to a cupboard, closed tightly, and there was a moldy odor coming from within. He shone a light through the crack and then stepped back, uncertain of what he had actually seen. It looked to him as if a naked woman were sitting inside that wall. He had seen her back.

In haste, Brown contacted the police, drawing Chief Superintendent Peter Beveridge to the flat, along with

Chief Inspector Percy Law of Scotland Yard. They pried open the door to reveal the decomposing corpse of a woman sitting amid some rubble. She wore only a garter belt and stockings, and her black sweater and white jacket had been pulled high around her neck. Behind her was something equally large, wrapped in a blanket that was knotted to the back of her bra. This corpse was removed and taken to the front room for examination, and it became clear as she was photographed that she had been strangled with a ligature. Her wrists were bound in front of her with a handkerchief twisted into a reef knot.

Next, the police focused on the other object in the cupboard. As they photographed it, they noticed another tall, wrapped object just beyond it. They pulled out the first one and soon discovered that it was another body. Oddly, it had been positioned head down. The blanket was fastened, also with a reef knot, around the ankles and the head was wrapped in a pillowcase.

They guessed before they saw it that the third object was yet another female corpse. This one, too, was upside down, with her head positioned beneath the second body. Her ankles were tied with an electrical cord, which was tightened into a reef knot. A cloth covered the head and was similarly knotted. The officers knew they had a significant case on their hands and they sent the bodies to the mortuary for a thorough examination. Other officers remained at the home to keep looking. They had noticed some loose floorboards in the parlor, so they pulled these up, dug through the loose rubble, and found yet another female body.

Meanwhile at the mortuary, the autopsies revealed more information. The first woman found had been dead about a month from strangulation and possible carbon monoxide poisoning. It was surmised that she had been under the effects of the poisoning when she was strangled with a smooth surface–type cord. She had been sexually assaulted at the time of death, or shortly after. Some four to eight weeks before her, the second victim had been strangled, sexually assaulted, and gassed as well. A white vest had been placed between her legs in a diaperlike fashion. The third woman, a blond, had been six months pregnant. She had been drinking shortly before death, which had taken place eight to twelve weeks earlier, also by asphyxiation.

The body from beneath the floorboards was an older woman, in her fifties. She had been dead three to four months. Unlike the others, there were no signs of coal-gas poisoning or sexual intercourse. She had been strangled, probably by ligature. She turned out to have been Ethel Christie, the wife of the former tenant. The others were prostitutes whom Christie had brought home to his near-empty flat: Hectorina McLennan, twenty-six; Kathleen Maloney, twenty-six; and Rita Nelson, twenty-five.

The police went through the entire flat, aware that a few years earlier, in 1949, a double murder had been committed at this address in an upstairs flat: the pregnant wife and infant daughter of Timothy Evans had been strangled, and Christie had been implicated but exonerated. Evans had confessed, then retracted his confession and accused Christie of a botched abortion, but nevertheless was found guilty of both murders and executed. They could only wonder, with these discoveries, if Christie had indeed been innocent. Further inside the kitchen cupboard, investigators found a man's tie, fashioned into a reef knot. In another area of the flat, they located potassium cyanide and a tobacco tin containing four clumps of pubic hair—three of which proved to be from the cupboard victims.

Once searchers had gone through the house, they focused on the back garden. In plain sight, they noticed a human femur supporting the wooden trellis. More bones and some hair and teeth were found in flower beds, while blackened skull bones with teeth and pieces of a dress turned up in a dustbin. Nearby was a newspaper fragment dated July 19, 1943. Investigators determined that, although only one skull was found, there were two sets of remains in the garden. That made a total of six victims, all female.

Anthropologists reconstructed the skeletons and, from a tooth crown in one skull determined that the woman had been from Germany or Austria. She was young, around twenty-one, and about five-foot-seven. The other was between thirty-two and thirty-five, and only about five-foot-two. Both had been in the garden at least three years and possibly as long as a decade.

Detectives eventually learned that Ruth Margarete Fuerst had arrived in England from Austria and had been missing since August 24, 1943. She was about five-foot-eight. The other victim seemed likely to be a Muriel Amelia Eady, thirty-two, who had worked at a factory with Christie. She was short and had dark hair, similar to hair from the garden. She had been wearing a black wool dress when last seen, and black wool was retrieved from the soil.

Mold was growing in the nostrils of the most recent victim found in the closet, and the last one to be removed was so encrusted with it that analyzing her time of death had been difficult with normal methods. Pathologist Francis Camps sent the mold samples to experts to learn their expected growth rates in a damp London cupboard. The biologists studied these samples and indicated that the mold had made a normal growth progression. They affirmed the initial hypotheses about the postmortem intervals.

Yet even before they delivered their results, Christie was found wandering the streets and he confessed to the murders. However, he said that his wife had died from an accidental overdose. Toxicologists proved that Christie's claim about Ethel was an outright fabrication; they also demonstrated the presence of carbon monoxide in the bodies of the three victims from the closet. In addition, this case showed medical professionals a heretofore unknown fact—that spermatozoa could be preserved for weeks in a corpse. Christie clearly had had sexual contact with the three younger victims.

Breaking down when confronted, Christie described how he had used a homemade contraption that dispensed a poisonous coal gas, filtered through Friar's Balsam. He would persuade the women that he could perform some medical service and assured them that the procedure required them to breathe the carbon monoxide gas. As they passed out, he raped them while strangling them and from some he collected pubic hair in a tin. He killed two before December 1952, placing them in the garden; then he murdered his wife before strangling three more victims. While in prison awaiting trial, he unashamedly sold his story to the *Sunday Pictorial*, including his part in the deaths of Tim Evans's wife and baby. This raised a controversy over the justice system hanging an innocent man.

Christie stood trial at the Old Bailey on June 22, 1953, on the charge of murdering his wife. The proceeding lasted four days. The presiding judge was Justice Finnemore and the prosecutor, Attorney General Sir Lionel Heald. Derek Curtis-Bennett defended Christie with an insanity plea. Psychiatrist Jack Abbott Hobson testified that Christie was a severe hysteric who may have known what he was doing at the time of each murder, but did not appreciate that it was wrong.

The prosecution had two distinguished professionals for rebuttal of this flimsy diagnosis, Dr. J. M. Matheson and Dr. Desmond Curran. Matheson said that Christie suffered from a hysterical personality, which was not a defect of reason, so he was not legally insane. Curran found Christie to be an "inadequate" personality with hysterical features, but with no defect of reason. Heald presented to the jury the account given by Christie of the murder of his own wife, which included evasive behaviors indicative of knowledge of wrongdoing. Thus, he had not been insane.

The defense attorney asked the jury to consider how abominable Christie's actions were and how revelatory of madness: a man who had intercourse with dying or dead women; a man who kept a collection of pubic hairs; a man living, eating, sleeping for weeks, even years, with those bodies nearby; he couldn't be sane.

The jury deliberated only an hour and twenty minutes before they found Christie guilty. He was sentenced to death and on July 15, he was hanged at Pentonville Prison.

That year, 1953, was groundbreaking for science. Biologists were already aware that DNA carried genetic information from one generation to the next, but no one had yet found its actual structure and hereditary mechanism. An American geneticist, James Watson, and an English physicist, Francis Crick, with Maurice Wilkins and Rosalind Franklin, worked many long hours together at the University of Cambridge in England until they were able to present their discovery: the double helical structure, which would become key to molecular biology and biotechnology. They had assembled information from other scientists working on various aspects of DNA to produce their model, relying on the knowledge that DNA was based in nucleotide subunits. Each nucleotide was comprised of a sugar (deoxyribose), phosphate, and one of four different bases—the purines, adenine (A), and guanine (G) together with the pyrimidines, thymine (T), and cytosine (C). Watson and Crick knew from work in the 1940s at Columbia University that these bases may occur in varying proportions in different organisms, but that the A and T residues remain in the same ratio to one another, as do the number of C and G residues. These relationships helped to establish DNA's three-dimensional structure and show how genetic information is encoded and passed on to successive generations. While this model was a breakthrough for science, it would not have ramifications for forensic science for three more decades.

In addition, during that decade at the University of Toronto, Dr. Robert Jervis pioneered applications of neutron activation analysis. First discovered in 1936 in Sweden by George de Hevesy and Hilde Levi, the process involved making items containing specific elements radioactive and thus identifiable via their nuclear signature. However, electronic equipment was needed, which left the application in a forensic investigation to those who had the full laboratory capability. That would occur later in the decade.

EVIDENCE ANALYSIS

In 1954, a Texas court admitted human bite-mark evidence in *Doyle* v. *State*. A grocery store had been burglarized and a piece of cheese was found on a counter, which showed the impression of teeth: The burglar had decided to grab a snack but left part of it behind. When a suspect was identified, the police asked him to bite into a similar piece of cheese, and he did. Photographs and casts were made, and a dentist affirmed that the same teeth had made the impressions on the two pieces of cheese. With this evidence and testimony, the suspect was convicted, and while his attorney appealed, it was based on the grounds that his rights had been violated rather than on challenging the bite-impression analysis. The court upheld the conviction.

Blood analysis figured into a frightening 1955–56 serial murder case in Germany. The "Düsseldorf Doubles Killer" managed three separate incidents, which always resulted in either double homicides or a murder and wounded survivor. The first incident involved two lovers who were battered and then left in their car as it was pushed or driven into a nearby pond. Two months later, two men were the victims, although one survived. A month went by and then another couple was battered and shot on February 7, 1956, and left in a burned-out haystack.

Investigators arrested and detained Erich von der Leyen, who had a reputation for behaving oddly and lived near the third crime scene. He had been home alone on the nights when each murder occurred, and police found spots in his Volkswagen that appeared to be blood. The Forensic Institute analyzed them with the Uhlenhuth method and said they were of human origin, as were stains on the suspect's clothing, from two different blood groups. The evidence was reexamined and more blood turned up on the suspect's trousers, but a careful analysis indicated that it was menstrual blood from a dog. Von der Leyen said that his girlfriend's dog had been in heat and had jumped into his car. When the Forensic Institute looked at their initial results, they were forced to admit their mistake and von der Leyen was released.

On June 6, a forester spotted a man with a gun stalking a couple in a parked car near Büderich. He arrested the man, who gave his name as Werner Boost, age twenty-eight, from Düsseldorf. At first Boost protested his innocence, claiming to be a married father of two. But eventually he admitted that he wanted to frighten couples having sex to make people around the country cleave to higher moral values.

Since the lone survivor of the three attacks had indicated that two men had gotten into the car with him and his lover, it was clear to the police that they should be looking for another suspect. In Boost's diary he'd mentioned the name "Lorbach" but the police did not need to search for him, as Franz Lorbach came in on his own to explain what he knew. He was not there to confess, however, but to turn in his partner. He said that Boost had led him into spying on courting couples, and then knocking them out with cyanide gas so they could rape the women. Lorbach insisted the one incident of murder in which he was involved had been an accident. Boost had killed one male and Lorbach had whispered to the other one to pretend to be dead, thereby saving him. Lorbach added that Boost had hypnotized him to get him to participate. While the police could not prove this last allegation one way or the other, they had enough to take both men to trial. They were convicted, and Boost was sentenced to life in prison, while Lorbach received six years.

From blood analysis to bones, the Trotter and Gleser formulae for stature estimation in human skeletal remains had been published by this time, reportedly used in the identification of Korean War dead. In 1957, American pathologists Thomas Mocker and Thomas Stewart identified skeletal growth stages, providing a basis for standardized measurements and analyses in forensic anthropology.

Also during the 1950s, a technique called neutron activation analysis, a method for testing metals, became a valuable forensic tool. Uranium and radium emit radiation rays of three types: alpha, beta, and gamma. Since the type and degree of energy emitted by a given element is unique, a scintillation counter can make a precise measurement, identifying the element via its radioactive signature. Substances not already radioactive can be made so by bombarding them with neutrons, or subatomic particles. It will give off gamma rays that can be identified via measuring their energy and intensity. This procedure was believed to assist with the identification of certain substances, specifically trace materials.

In an investigation, then, hair, soil, glass, or paint could be placed inside the core of a nuclear reactor and bombarded. The neutrons would collide with the components of the trace elements in that substance, changing them radioactively to emit gamma radiation of a characteristic energy level. With this method, a scientist could measure the sample's constituent parts, no matter how small. In a single hair, for example, fourteen different elements could be identified, allowing for fairly precise comparisons between hair from an unknown person and hair from an identified suspect, based on composition.

Whether the metal is organic or inorganic has no effect; either could be analyzed effectively with this process.

The first case to admit neutron activation analysis in court occurred in 1958 in Canada when John Vollman was prosecuted for the murder of sixteen-year-old Gaetane Bouchard, whose body had been found near Edmunston, New Brunswick. In her hand were a few strands of hair. Vollman lived in Maine, across the border, but the girl's father was aware that he had been seeing Gaetane. In fact, witnesses placed the two together just before she was discovered dead. Mr. Bouchard went to Vollman's home to ask about Gaetane, but Vollman said that they had broken up. He did not know how she had been killed or who might have done it.

Yet the physical evidence told a different story. A chip of green paint picked up by an observant police officer from the murder site appeared to match the color of Vollman's car. When it was searched, a partial piece of candy was found in the glove compartment that bore a trace of lipstick the same color as that which Gaetane wore when she died. Indeed, the hair strands clutched in her hand were the same color as Vollman's hair. Given these links, investigators asked for another examination of her body. This time, the pathologist was more thorough and a single strand of hair was found wrapped tightly around one of her fingers. It, too, was the color of Vollman's hair.

They subjected this strand to neutron activation analysis, and did the same to samples of Gaetane's and Vollman's hair, which revealed that the ratio of sulphur radiation to phosporus was closer to his than to hers. Once this evidence was admitted, Vollman changed his plea to guilty of manslaughter. The jury instead found him guilty of murder.

By the end of the 1950s, forensic science had become established and was now in the process of innovation without the various fields. This involved creative thinking and talented professionals to devise new methods, as well as refine older ones (or realize their limited value). While professionals still disagreed in court over how to interpret tests and evidence, the participants in the legal arena had come to accept science as a trustworthy, if somewhat imperfect, complement to their efforts. Hereafter, scientists and technicians experienced less resistance to their presence in the courtroom, but they were still held accountable by their colleagues and by opposing attorneys for what they might offer in testimony. Without the need to prove the merit of science itself, they were free to experiment a little more and to seek different ways to apply science than had been done before. Sometimes that proved to be wonderfully innovative and sometimes a little silly.

OUTSIDE THE BOX

Knowledge of plant life peculiar to certain areas can offer important information for making decisions in criminal investigation. In June 1960, Bazil Thorne in Sydney, Australia, won a lottery, which garnered a fair amount of publicity. He had an eight-year-old son, Graeme, who was soon kidnapped on his way to school. Within an hour, the Thorne family received a call from a man with a thick accent demanding a one-quarter share of the lottery money. If it was not paid by that afternoon and done according to his directions, he stated, the boy would be fed to the sharks. Mrs. Thorne recalled a man who had come to her door a few weeks earlier asking about a family who had never lived there, and this report provided the police with a description of a viable suspect.

On July 8, the boy's emptied school bag was found, and nearby, its contents had been scattered along the roadway. Over a month passed with no sign of Graeme, but then on August 18, his fully clothed body was discovered lying on a vacant parcel of land in Seaforth. He had been gagged and bound, wrapped in a tartan blanket and stuffed under some bushes. An autopsy showed that he had been bludgeoned with a blunt weapon. Detectives went over the blanket, looking for trace evidence. They found hair from a Pekinese dog, soil, pink limestone mortar, and seed from two rare types of cypress tree that did not grow near the body dump site. The police also knew that a blue Ford had been seen on the morning of Graeme's abduction, so they canvassed the area, looking for cypress trees, a car, a house with pink mortar, and a man with a thick accent. It took about six weeks, but they finally found what they were looking for.

A house belonging to a Steven Bradley was made from dark brick, which had been cemented with pink mortar, and in the yard were the right type of cypress trees. The

Bradleys had also once owned a Pekinese dog, which was no longer there. Inside their blue Ford, investigators found a dog brush with hair similar to that on the blanket used to wrap the victim's body. However, to their chagrin, the house was empty and they learned that Bradley was gone from the country. They did acquire a photo, and Mrs. Thorne looked at it and identified Bradley as the man who had come to her home. In addition, pictures developed from a discarded roll of film revealed that Bradley had once owned a tartan blanket. He'd grown up in Hungary, so he had an accent.

Certain that Bradley was the kidnapper/murderer, the police located Bradley through various records and flew him back to Australia. He claimed that while he had abducted the child, the boy had died accidentally. However, this story failed to hold up, and Bradley was found guilty of murder and sentenced to life in prison.

Another case in England likewise provided the opportunity to try an untested technique. In London's Cecil Court, near Charing Cross Road, Edwin Bush, twenty-one, entered an antiques shop on March 2, 1961, and asked to examine a dress sword and some daggers. The next day, a young apprentice discovered the body of a woman in the shop, stabbed deeply three times. A dagger was still sticking out of her chest. Her husband identified her as Elsie Batten, an employee of the store.

Detective Sergeant Raymond Frederick Dagg interviewed the shop owner, Louis Meier, and he gave a description of the young man who had asked to see the daggers. Dagg used the opportunity to compile the features from the Identikit, an instrument in which he had just received training. He confirmed it with another interview in a shop down the road where the same man seemed to have stopped. In fact, he built two likenesses from the two descriptions, and they were considered so similar that Dagg's supervisor believed it had to be the same man. A photograph was taken and released to the media. Less than a week after the murder, a police officer spotted Bush, identifying him from the facial composite. He arrested the young man, who was then placed in a lineup. One of the witnesses identified him. Bush then confessed, saying he did not know why he had murdered the woman. Sentenced to death, he was executed.

In a more peculiar case in 1961 in England, David Cohen, a branch secretary for the Society for Psychical Research, approached Detective Chief Inspector Tony Fletcher, who worked at the Manchester Police Fingerprint Bureau, with an unusual request. Cohen wanted to know if it was possible to fingerprint a "nonphysical entity," and if so, would the police assist him in doing so? Fletcher, who later recorded the tale in a memoir, thought it was an interesting question, so he sent an officer to see what could be done. Supposedly, the ghost of a musician, an elderly man named Nicholas, was playing a violin in the room of a young boy in a south Manchester home. Séances were held at the house under Cohen's guidance and at times he had noted a pair of spirit hands appearing. He believed that if it were a trick, fingerprints would reveal the perpetrator. If not, then fingerprints taken in such a circumstance would make news around the world.

The officer apparently saw the hands for himself and described them as pale and slim, with lace cuffs. He dutifully polished a tambourine that appeared to fly about the room, and after the séance used mercury powder to test it. But there were no prints.

The next experiment involved putting the powder on before the séance, yet still nothing showed up, although the instrument performed its usual act. The group then addressed the entity, asking for and receiving permission to take its fingerprints on chemically treated paper. By all reports, it agreed. The officer said later that he even felt the hand in his own as he pressed it to the paper, but still no fingerprints materialized. Only three scratches showed up where the fingers had been placed. They also tried infrared photography, which yielded a faint image but nothing definitive. Stymied, they could think of no other methods to record this entity, and eventually the experiment was abandoned.

Few scientists are even aware of it, and probably none would view it as an appropriate use of science, yet if not for people who tried to see connections where no one else had before, some of the following discoveries might never have been made.

ELEVEN

PRACTICAL SCIENCE

MAKING SURE

President John F. Kennedy was assassinated in Dallas, Texas, on November 22, 1963, while riding in a black open-top Lincoln Continental before an exuberant crowd. A bullet pierced through him to wound Governor Connally, seated in front of him, while a second one slammed into his skull. Kennedy was rushed to Parkland Memorial Hospital for emergency treatment, but after he died he was transported to Bethesda Naval Hospital just outside of Washington, D.C., for an autopsy. Quickly classified reports soon inspired theories about an assassination conspiracy and a government cover-up. Lee Harvey Oswald was arrested for the shooting, but Jack Ruby emerged while Oswald was being transferred to a jail and shot him, some said to silence him. Although forensic specialists had no access to Kennedy or the assassination documents, persistent rumors, along with the demands of forensic pathologist and Pittsburgh-based coroner Cyril Wecht, would eventually force the issue.

In the meantime, in England, entomology helped solve another case. In June 1964, two boys found the partially buried body of a man in the woods outside Bracknell. Keith Simpson was among those who went to the shallow grave to examine the remains. The man's head had been wrapped in a towel, so he clearly had not died in the woods from some accident. When the police attempted to estimate his time of death, decomposition had reached the point where it was difficult to make an assessment. However, Simpson had studied the behavior of blue-bottle fly maggots and from his calculations of their predictable stages of development, he believed that the man had been in this place between nine and twelve days, closer to the former than the latter. He later wrote about the case in his book *Forty Years of Murder*. An autopsy revealed that the victim had died from a strong blow to the throat, and from police reports the discovered remains were linked to area resident Peter Thomas, missing for the past ten days.

The police learned that Thomas had loaned money to a William Brittle, which made Brittle a suspect. Yet he insisted that he'd repaid Thomas on June 16, and when a witness claimed that he'd seen Thomas alive four days later, the case weakened. Nevertheless, Brittle was arrested and tried. This proceeding provided a forum for Simpson to demonstrate what he knew about maggots and decomposing corpses, which clearly proved that

Thomas could not have been alive on June 20, as per the witness's statement. Even the entomologist for the defense conceded the truth of Simpson's findings, so Brittle was convicted and sent to prison. Simpson used this same technique in other cases, establishing entomology as a viable science for estimates of the time that has elapsed since a death occurred.

In 1965, Oatley McMullan and Ken Smith invented the first high-resolution electron microscope, signaling how forensic investigation was becoming more involved with precision technology. Some of these advances assisted in revisiting cases once thought finished, for sophisticated reconstruction or better analysis.

One from 1959 was reopened in Canada in 1966 as a near-crusade. After dinner on June 9, twelve-year-old Lynne Harper left her home on the Clinton air base in Ontario, to take a walk. She encountered an older classmate, Steven Truscott, fourteen. He invited her to go for a ride, so she climbed onto the crossbar of his bike. Another child reported seeing them around 7:10 that evening on a road north of the base. A few minutes later, a boy riding his bike on the same road claimed that he'd seen no one, and someone else reported that he'd seen Steven alone. It seems that the first child was the last one to have seen Lynne Harper alive.

She failed to come home, so her parents called the police, who organized a search. No one found her that night, but the police continued the search the next day. They already knew that Truscott had been with her, but he claimed to have dropped her off in a specific spot and had seen her get into a car. That had been around 7:30. He then went to play a sport with friends and arrived home about an hour later. When asked for the car's description, he said it had been a new Chevrolet, gray, with a yellow license plate. Since Michigan cars, just across the border, had such plates, it sounded as if he were telling the truth. He had not seen the driver so he could provide no description.

It was two days before Lynn's body turned up in the woods. She'd been sexually molested and then strangled with her blouse. Her shoes and socks were laid out in a neat arrangement, along with her shorts. Her panties lay thirty feet away. Nearby were the tire tracks from a bicycle and the footprints of a person wearing crepe-soled shoes. At the scene, Dr. John Penistan turned the body onto its left side to photograph the ground. To estimate her time of death, which he said could no longer be attained with rigor mortis or liver mortis measures, Penistan analyzed the contents of her stomach and the state of digestion of the meal she had eaten just before leaving home. He estimated that she had died around 7:30 P.M. that same evening. Thus, she had been killed around the time she'd been with Steven Truscott, less than two hours after leaving her home.

Investigators determined that from the vantage point that Truscott said he was at, he could not have caught a clear glimpse of the car he claimed to have seen. They knew it was time to subject the boy to more rigorous analysis. Suspicion deepened when the tread on his bicycle tire matched the impression left at the crime scene. A physician found marks on his penis consistent with rape, although Steven protested that they were from a rash. However, he did own a pair of crepe-soled shoes, which he was unable to produce after a search indicated that they were no longer among his effects. In addition, the semen removed from the victim revealed a secretor and matched Truscott's blood type, but since Lynne had the same blood type, this evidence proved less compelling. Still, Truscott was charged with murder.

The strong circumstantial evidence, along with testimony from a boy whom Truscott had asked to manufacture him an alibi, ensured a conviction, and Truscott was sentenced to hang. However, due in part to public outrage over killing a child, his sentence was commuted to life. Eventually an appeal was filed to Canada's Supreme Court, which ordered a judicial review. A journalist picked up the story and viewed the evidence as favoring Truscott's innocence, claiming in her bestseller *The Trial of Steven Truscott* that there had been an unfair ruling in the case. She hoped to get the attention of two prominent pathologists in Britain, Keith Simpson and Francis Camps. They both read her argument, and while Camps agreed and claimed he wanted to help see justice done, Simpson inclined toward the findings of the original pathologist. They ended up in Canada in 1966, on opposite sides.

Experts also came from the States, aligning themelves in one camp or the other, with the focus centered on the analysis of the estimated time of Lynne Harper's death. If Penistan had been wrong, that would provide reasonable doubt for Truscott: If Lynn had in fact died an hour or more later, then Truscott had an alibi and could not have been her killer. New York pathologist Milton Helpern, with decades of experience, affirmed the time of death at approximately two hours after Lynne had left home.

The prosecution was prepared to go before a nine-judge tribunal to argue the case again, hoping to avoid the inconvenience and expense of a retrial. Every piece of the original evidence had been preserved, and experiments with blouses like the one Lynne had worn proved that she could have been murdered with it (which the journalist claimed was impossible). Another issue was whether marks on the girl's face proved that she had been moved after she was killed to where she was found, but the prosecution was able to show that Penistan's moving the body at the crime scene for photographs of the ground had caused them.

Since the sole time-of-death indicator had been the process of digestion, Simpson and Camps debated its accuracy. It had been determined that she had eaten at 5:30 P.M., and it had seemed impossible to Penistan from the amount of food remaining in her stomach when she was found that more than two hours had passed. Because there are variations in the digestive process, depending on the type of food eaten and the speed of the person's metabolism, the experts siding with the defense argued that the assessment could easily have been wrong. In addition, they added, certain strong emotional states such as terror or anger can also slow the process, and the girl had certainly been afraid, as she was being assaulted.

In a surprising statement, Camps claimed that Lynne could have died as long as ten hours after her last meal. Countering this, Simpson pointed out that the death scene supported the initial analysis. In addition, no suspicious gray Chevrolet had been found and there were no clues pointing to other suspects. The time of death estimate, said Simpson in his calm and methodical manner, was more likely to be correct than as inaccurate as Camps was claiming.

In the end, the judge decided that Camps and his allies had provided no reason for a new trial (and Camps even conceded that two hours was a possible time frame), so the sentence was reaffirmed. Yet Truscott was released on parole in 1969, ten years after the murder, as if sufficient doubt had indeed been raised to just let him go.

Just south of Canada, in Ohio, Sam Sheppard, an osteopath from Bay Village, also appealed his case in 1966, with much more emphasis on evidence analysis. It was also a much older case.

Early on July 4, 1954, Sheppard had phoned the mayor (his neighbor and friend) and then the police to report that his pregnant wife, Marilyn, had been murdered in their upstairs bedroom. She had been brutally bludgeoned, they saw, and Sheppard told a strange story of how he'd been asleep on a daybed at the foot of the stairs. Apparently the intruder had crept past him and up the steps, for he'd awakened to the sound of Marilyn screaming for him. (Their seven-year-old son sleeping in the next room did not wake up.) Sheppard had then encountered this intruder, a "bushy-haired stranger," and they had struggled twice, but Sheppard was knocked out (twice) and the man escaped. Sheppard had few injuries from this encounter, all of which were documented at the hospital that his family operated, so no independent doctor had examined him. He'd described getting blood on him when he'd checked to see if Marilyn was alive, but oddly, there was no blood on his hands. Yet there was blood on his watch, which would later become an issue.

There were also blood spatters throughout the room where Marilyn lay, blood smears on the sheets, and a blood trail leading through the house. Multiple gashes scored her face, as if her killer had been enraged. On the first floor, Sheppard's medical bag was overturned, but had no blood on it, and a desk was rifled, also without the killer leaving blood behind. Sheppard's story sounded fishy, so the police arrested him. It became clear as they questioned him, and later at his trial that October, that his account had glaring holes, including lies about an extramarital affair and his seeming inability to tell the story straight.

Yet both sides failed in a number of significant ways to prove their cases. No one examined Marilyn for evidence of rape, for example, or compared Sheppard's blood type against the blood trail found in the house. Although the prosecution managed to give Sheppard a viable motive with testimony that he had intended to file for divorce, they were unable to produce the murder weapon. Their medical expert insisted it had been some type of surgical instrument, but he was unable to identify it. Not surprisingly, the jury took the middle ground and returned a verdict of second-degree murder. Sheppard received life in prison.

A renowned criminalist, Dr. Paul Leland Kirk, was invited to assist with evidence for an appeal. Kirk was head of the criminalistics department at the University of California and in 1953 had published *Crime Investigation: Physical Evidence and the Police Laboratory,* which would become a classic textbook for forensic science. He coined the term *blood dynamics*, for the scientific approach to bloodstain pattern analysis.

Kirk had examined the crime scene and autopsy photos, with special attention to the blood spatter in the room. He testified that the killer had been left-handed, which would exclude the right-handed Sheppard. He also said that the blood spatter on Sam's watch was a transfer stain, which had probably come from contact with Marilyn's body when he tried to find her pulse. Kirk had examined an additional spot of blood from the bedroom that he said had originated with someone other than Sam and Marilyn. While it was type O, the same as Marilyn's, it had other components (observed from an unrepeatable test) that gave it a different composition. However, even with the weight of Kirk's credentials and analysis, the appeal failed. Nevertheless, Kirk's findings became fodder for several other theories, which generated books denouncing the original investigation.

In 1959, a window washer named Richard Eberling was arrested, and among his effects was one of Marilyn's rings, taken from the home of her sister-in-law. He admitted that he'd washed windows for the Sheppards not long before the murder and offered information, unsolicited, that he'd cut himself and dripped blood on the basement steps. So he was now a suspect, but it would be four decades before this information became a serious part of the case.

Defense attorney F. Lee Bailey took over, filing a motion with the U.S. District Court to overturn the conviction based on prejudicial pre-trial publicity. The court acknowledged this, but its decision was reversed. Then in a precedent-setting decision in 1966, the U. S. Supreme Court ruled that Sheppard should get a second trial.

A book had come out in 1961 proposing that Sheppard was innocent and that a man and woman had killed Marilyn together. Along with Kirk's blood analysis, this gave F. Lee Bailey what he needed to propose a new theory. His version honed in on two people: the neighbors whom Sheppard had called for help right after the murder, and Sheppard's story about encountering a stranger was not even mentioned. Nor did he testify. Bailey believed that a left-handed woman had struck Marilyn and had been bitten during the attack, so she had left the blood trail. However, he was not allowed to enter much evidence, so he resorted to embarrassing the prosecution's witnesses on cross-examination.

This second jury returned a verdict of not guilty. Sam was a free man, but he became an alcoholic. When his anemic medical practice went bust, he looked into professional wrestling, but by age forty-six he was dead from liver disease. When his son attempted a civil suit against the state years later to prove with DNA that Richard Eberling had been the killer and that his father had been unlawfully imprisoned, he lost. His expert had indicated that the DNA pattern could not rule Eberling out, but a basic blood test that the defense performed contradicted his findings. By this time, Eberling, who was right-handed, was dead. So were the Houks, whom Bailey presented once again as suspects during this civil proceeding. Some people believe the case remains unresolved, and books continue to be written to offer new theories.

VIOLENT DECADE

Five years after Kennedy's assassination it was his brother's turn: Robert Kennedy was gunned down during his campaign to become the next president. Martin Luther King, Jr., was also assassinated during this time. Around the United States, university students demonstrated against the war in Vietnam and blacks rioted in a show of anger over violations of their civil rights. In addition, the number of murders by strangers had increased, and killers with monikers such as the Boston Strangler, the Co-ed Killer, and the Pied Piper of Tucson were grabbing headlines. In 1967, the FBI started the National Crime Information Center (NCIC) to coordinate investigations across jurisdictions. This involved a computerized index that permitted state and local jurisdictions access to FBI archives on such items as license plate numbers and recovered guns, as well as the ability to post notices about wanted or missing persons.

Another bite-mark case that year gained prominence. In Biggar, Scotland, fifteen-year-old Linda Peacock had gone missing. Officials searched all night before they discovered her body in the local cemetery. She had been strangled and beaten, and her bra and blouse were in disarray. On her right breast was an odd bruise. Bite marks had not yet been used in a British court for the definitive identification of a perpetrator, although their worth for identification had long been recognized. This case would set an important precedent.

Since the bruise appeared to have been made during the struggle, investigators took numerous photographs. Analysis indicated that whoever had killed Peacock had bitten her hard. Dr. Warren Harvey, a forensic odontologist, confirmed that the bruise was indeed a bite mark and upon closer examination indicated that there was some unevenness to the killer's teeth. They looked strangely jagged.

It seems that there were also witnesses to the crime. A male and female had been spotted at the cemetery gates the night before, and the girl resembled Peacock. From the way these two had addressed each other, they seemed familiar. The witness had spotted them around ten o'clock in the evening and about twenty minutes later, this same person heard a girl screaming.

A systematic search of people's teeth was undertaken to try to eliminate area residents, and it included thirty inmates at a detention center. Everyone was asked to provide dental impressions to compare to the victim's bruise. Dr. Harvey studied them all and narrowed the suspects to five, asking each for another impression. At this point, pathologist Keith Simpson joined the team. Together these men examined all the impressions and agreed on a single suspect: seventeen-year-old Gordon Hay, arrested for breaking into a factory.

When the dentist took this second impression, he found that one of Hay's teeth was pitted in two places by hypocalcination, and the pit locations matched the odd marks on the impressions from the victim's breast. Harvey was confident he could prove the match in court, even though it would be the first time that this type of testimony would be used as the defining evidence.

To strengthen his presentation, Harvey examined the teeth of 342 young soldiers. Only two had pits of any kind, and none had the particular two pits that shaped Hay's teeth. From this analysis, Harvey concluded that Hay's teeth were so unique that it would be virtually impossible to find another set of teeth like his that could come as close to the bruise impression.

At his 1968 trial, Hay claimed he was at the detention center at the time of the girl's death, so he could not be the person they were looking for. However, another inmate stated that Hay had actually come into the dormitory later than he told the court and there had been mud on his clothes. Another witness said that Hay had met Linda Peacock at a fair just before she was murdered, and he had confided to friends that he planned to have sex with her.

To take the case beyond circumstantial evidence, the prosecution introduced the dental results. Since this type of analysis was so unique, the defense team fought to have it ruled inadmissible. When the judge allowed it, the attorneys then brought in their own dental expert to refute it, or at least to confuse the issue. The jury apparently accepted the quality of the evidence because Hay was convicted of murder. Still, the defense did not give up; they appealed on the same basis. However, the court upheld the judgment, which set a precedent for similar cases with this type of evidence.

The first Canadian case to identify a perpetrator from bite marks in skin began that same year, 1968, and involved a series of fatal attacks. A young schoolteacher, Norma Vaillancourt, was found murdered in her Montreal apartment. She'd been strangled, raped, and bitten all over her breasts. The crime was sadistic, but from among her many boyfriends there were no good suspects. Only a day later, another victim was found in that town in the same condition, and the bite marks linked the cases. In both incidents, there appeared to have been little struggle, so it was assumed that the women may have been engaged in something they wanted to do. Then Marielle Archambault told coworkers that she felt entranced by a man she'd recently met. She, too, turned up dead.

There was another victim across the country in Calgary before the "vampire" was stopped in 1971. Schoolteacher Elizabeth Ann Porteous was found lying on the floor in her apartment on May 20, with several bite marks on her breasts and neck. The suspect whom the police identified was a traveling salesman, Wayne Clifford Boden, and while he was being investigated, the police invited odontologist Gordon C. Swann to take casts of Boden's teeth to compare to the victim's bite marks. He used a system of geometric progression and found twenty-nine points of similarity. He believed that Boden had made the bite marks. The chief justice agreed, citing this evidence when he sentenced Boden to life in prison.

Dr. Swann also compared Boden's teeth to the marks on two of the three victims from Montreal (one had no bite marks). In the case of one victim, the photography had been poorly done so Swann was unable to make a determination. He did match the bite marks from Shirley Audette from Montreal to the Calgary victim and then to Boden, with seventeen points of similarity. Boden went to Montreal for trial, where he pled guilty to those three murders, claiming they had occurred during rough sex. He received three more life sentences.

Around the same time, in Britain, photographer Jacques Penry refined the Identikit system into the Penry Facial Identification Technique, or Photo-FIT. The images came from photographs instead of drawings for a closer approximation of what people actually looked like. Penry claimed that his approach dated back to 1938 when he was illustrating a book, *Character from the Face*. The home office hired him in 1968 to develop a kit, and his first one could theoretically build as many as five billion Caucasian faces. He added a supplement for black and Asian features, then devised another supplement for female identifications.

Witnesses could choose from a series of foreheads, mouths, hairlines, eyes, noses, and chins, along with headwear, facial hair, eyeglasses, and other items. The final result boosted the kit's power to about fifteen billion possible combinations. This form of composite identification became the precursor to the computerized images in use today.

VOICES

Voiceprint technology had received limited notice for criminal investigations by the early 1960s when the New York City Police Department received numerous bomb threats by phone against major airlines. Stymied, the FBI had requested assistance from Bell Labs. Lawrence G. Kersta, a senior engineer who was experienced with the sound spectrograph and had developed the voiceprint in 1941, acquired the task of devising a method of identification that would stop the calls and bring the perpetrators to justice. It took him more than two years and the analysis of over fifty thousand voices, but he managed to offer a technique that he claimed tested at a rate of accuracy over 99 percent. While he did not solve the mystery of who was making the calls, Kersta eventually broke away from Bell Laboratories to market the machine on his own, and law enforcement invited him to develop applications.

In 1971, the Wisconsin police used voice recognition

during the investigation of the September 24 murder of game warden Neil LaFeve. That afternoon, he had been out in the woods posting signs and had planned to finish long before the party that his wife had organized for his birthday. When he failed to show up, she grew worried and phoned his boss. They discussed it together, but there was no reason they could think of that Neil might still be in the woods. LaFeve's boss drove out to have a look. He noticed that all the signs were posted, so he called the police. Their search that night came up empty, but in the morning they located LaFeve's truck with the door ajar. Not far away on the ground was a considerable amount of blood, a pair of broken sunglasses, and two spent shells from a .22 rifle. In fact, they were able to follow a trail of items, from teeth to blood to bone fragments, until they came across an area of freshly dug earth. With shovels, they soon located Neil LaFeve, but his head was missing. Another soft spot nearby yielded it. His head had been hacked off with a blunt instrument, and he'd also been shot several times.

Looking for someone with a vendetta, detectives scoured a list of men that LaFeve had arrested for poaching. All who had been convicted of hunting illegally on those grounds were located and interviewed on tape, and a few were asked to submit to polygraph examinations. However, there was one man who refused to cooperate: twenty-one-year-old Brian Hussong. LaFeve had arrested him several times, yet he had continued to poach. Hussong had no alibi for the day in question and when he resisted attempts to clear up the murder mystery, he seemed a good suspect.

Sergeant Marvin Gerlikovski managed to acquire a court order that allowed him to put a wiretap on Hussong's house. He took the extra precaution of recording everything that was said, which paid off in a way he didn't expect. At one point, Hussong called his grandmother to urge her to hide his guns and provide an alibi. She appeared to cooperate, so Gerlikovski sent detectives to her house. Flustered, she led them straight to the hiding place. Ballistics experts confirmed a match between the .22 rifle and the bullets found in LaFeve's body, which was sufficient evidence to place Hussong under arrest.

Gerlikovski then sent the tapes he had made to Michigan's Voice Identification Unit—at that time the best in the world for voiceprint analysis. Ernest Nash examined the tapes, gave his opinion, and served as an expert witness during Hussong's trial. However, it was not Hussong's voice that he testified about, but that of Hussong's grandmother. She had denied saying that she had hidden the guns, so Nash explained how he could match her voice to that of the voice on the tape. He then used his laboratory results to affirm that she was definitely the person speaking on the tape. That meant that Hussong, the man talking with her, was worried about the guns. The jury listened to the tapes for themselves and returned a guilty verdict of first-degree murder, which resulted in a life sentence for Hussong.

LARGER DEVELOPMENTS

Since 1972, fingerprints had been retrievable via computer. The FBI put into motion a plan for the Automated Fingerprint Identification System (AFIS), which was finally established in 1975. The following year, Scotland Yard utilized a national fingerprint computer called Videofile, containing more than two and a half million prints from criminals. Just as significant was the inclusion of the gas chromatograph-mass spectrometer (GC-MS) in forensic use. Chromatography was a method by which compounds could be separated into their purest elements, as inert gas propelled a heated substance through a glass tube where a detector charted each element's unique speed for a composite profile. The mass spectrometer, linked to it, affirmed the identifications via patterns of spectra. With this equipment, forensic scientists could analyze such items as hair samples for drugs or poison, charred remains for accelerants, and the chemical composition of explosives.

During this time, the FBI also established the Behavioral Science Unit (BSU) at their training academy on the marine base at Quantico, Virginia, for a more effective approach to the investigation of serial crimes. By 1977, the BSU specialized in crime scene analysis, criminal profiling, and the analysis of threatening letters. As murder rates rose during the 1950s and 1960s the FBI had received expanded jurisdiction, and a handful of agents versed in abnormal behavior had devised a way to "profile" behavioral evidence from crime scenes. Special Agent Howard Teten had met psychiatrist James Brussel during the 1960s to learn how he had accurately provided a personality composite of Manhattan's Mad Bomber. Teten added what he knew from criminal investigation and with Special Agent Patrick Mullany, refined the approach for law enforcement. He then trained other agents to become crime consultants.

The BSU started with eleven agents, who taught and offered advice to local law enforcement on different types of crimes. Pattern violence quickly became their forte, and they expanded their skills from teaching to investigation, developing an identity as the Crime Analysis and Criminal Personality Profiling Program. Once these agents gained national recognition from their investigation of high-profile serial murders, the unit became firmly entrenched.

Japan made a significant contribution during the 1970s as well. Fuseo Matsumur at the Saga Prefectural Crime Laboratory at the National Police Agency observed his fingerprint on a glass slide and mentioned it to colleague Masato Soba. A few months later, Soba developed latent prints on smooth surfaces with Superglue fuming.

Around the same time, another detection method was developed in England when Bob Freeman and Doug Foster invented the electrostatic detection apparatus. This device, too, had been intended for detecting fingerprints on paper, but indentations from writing corrupted the readings. But then they spotted another application: The apparatus could be used for handwriting analysis—

those very indentations that got in the way. The method involved pinning a sheet of writing-indented paper between a glass plate and a sheet of Mylar. It was then placed on a machine with a brass plate and a lid charged with high voltage. A fine electrostatic substance sprinkled onto the paper moved toward the brass plate, filling in the paper's indentations to reveal a clear presentation of the handwriting. The resulting image could then be photographed. While this was not an advance in the actual art of analysis, it made the process easier in those instances where the original note was missing but the tablet remained.

CONSPIRACY

Issues with the Kennedy assassination resurfaced in 1972 when the *New York Times* announced that Kennedy's brain was missing. This revelation was the result of work undertaken by Dr. Cyril Wecht, a forensic pathologist, when he examined the Warren Report on behalf of the American Academy of Forensic Sciences. He was the first nongovernment forensic pathologist permitted to observe and study the autopsy materials, which had been preserved in the National Archives. He thought that the handling of the former president's body had been appalling, because the autopsy, performed by a pathologist with no experience in gunshot wounds, had been superficial and the medical photography poorly done. He was also disturbed by the notion that a single bullet could have done as much damage as he saw in the autopsy reports. He concluded that the Warren Commission's "lone assassin, single-bullet" theory could not be supported. Further, Wecht was dismayed to discover that certain key items were missing from the collection: photographs of Kennedy's internal chest wounds, glass slides of his skin wound, and most important, his brain. Wecht fed this information to the *Times*, hoping to bring public pressure to bear on the government to get a better investigation.

In 1978, Congress appointed a subcommittee to look into the matter, led by New York City medical examiner Michael Baden, who recruited Wecht and seven other medical examiners to assist. Baden and Wecht both wrote accounts of their experience and findings. One goal was to examine the possibility via trajectory wounds that there had been more shooters than just Oswald. However, Baden's team soon discovered they were at a disadvantage. As they examined the crime scene and autopsy photographs, Kennedy's clothing, the x-rays, and other reports, it soon became clear that the people in charge in 1963 had not realized the difference between a forensic autopsy and a regular autopsy. For example, Commander James J. Humes, the physician who performed the autopsy, had apparently not recognized the difference between an exit and entrance wound, and therefore could not pinpoint the bullet's origin. He also couldn't tell how many shots had hit the president.

The Secret Service had instructed Humes not to perform a complete autopsy, but only to find the bullet, believed to still be lodged in the body. In his reports,

Hume's medical descriptions were negligible, and he referred interested parties to the photos, which were so poorly focused by the photographer they were nearly worthless for medical purposes. Humes didn't even turn Kennedy over to look at the wound in the back of his neck, or call the receiving hospital in Dallas to discover that a tracheostomy had been performed, which he'd have found going right through the exit wound in the throat. He erroneously assumed the bullet had fallen out the same hole it had entered. He also failed to shave the head wound to see it clearly, and he had it photographed through hair. In addition, Humes miscalculated its location by an error of four inches.

After only two hours (a very short time), he prepared the body for embalming. Then, because his notes were stained with blood, he burned them. After he found out about the procedure done in Dallas, he rewrote his notes based on what he recalled and what he could figure out. He ended up including material he himself never saw and failing to track the bullets properly. Thus, his report was filled with errors.

By relying on the placement of bullet holes through the clothing and their experience with exit and entrance wounds, Baden's team managed to piece together the fact that two bullets had entered Kennedy. There was a small hole in the back of Kennedy's shirt and jacket, and small exit holes going through his shirt collar and tie. That was the bullet that had pierced his throat and gone into Governor Connally (having fallen from Connally's leg while he was on a stretcher). The other bullet had gone through the back of Kennedy's head and came out over his right eye, ending up hitting the car's windshield post and falling to the floor. Both had come from behind.

With the exception of Wecht, most of the pathologists on the Select Committee agreed that these two bullets had caused the wounds to both Kennedy and Connally, and that a single bullet had gone through Kennedy and into Connally. Wecht stood firm on his idea that Kennedy was struck twice in a synchronized fashion, from the rear and the right front side. He noted that before Oswald died, he had been interrogated for two days by top experts, but none had thought to keep any notes or to record the proceedings. It seemed to him unlikely that there was not a single piece of written documentation of one of the most important interrogations in American history. Despite the extensive analysis these pathologists performed, the conspiracy theories have not been laid to rest, and the longer the public is made to wait for the unveiling of documentation, the more firm the belief that there's something to hide.

ONE MORE BITE

Another bite-mark case became a media sensation in the States. Notorious serial killer Theodore Robert Bundy was intent on defending himself in court. Having been a law student, and assured that his own confidence, charm, and good looks could win a jury, he arrogantly prepared his case. His first trial in Florida began in July 1979, centering on two murders that had occurred on January 15, 1978. Lisa Levy, twenty, and Martha Bowman, twenty-one, were in bed in the Chi Omega sorority house at Tallahassee's Florida State University. A man wearing a blue knit cap crept in and struck them with a wooden log until they were dead. A sorority sister, Nita Neary, saw the man run from the house, so she called the police.

Levy had been raped, strangled, sodomized with a bottle, and beaten on the head, while Bowman had been severely beaten and strangled with a pair of panty hose. Two other girls in the house had been attacked and less than an hour and a half later, the man assaulted a fifth victim a few blocks away, who survived. Yet there were no leads, no fingerprints, nothing left behind, aside from semen and an odd bruise on the buttocks of Lisa Levy. One officer laid a yellow ruler against the abrasion and then stepped back for the photographers. His presence of mind might have made all the difference between conviction and acquittal of the most notorious killer in America, because by the time the case went to trial the tissue specimens had been lost.

Only a few weeks later on February 9, the killer continued to expose himself. Driving a stolen van, he abducted twelve-year-old Kimberly Leach in the middle of the day from near her school. He took her to a wooded area in a state park, where he raped and killed her, leaving her body there. Less than a week later, on February 15, he was driving a stolen Volkswagen in Pensacola, and because he was driving quite slowly the police ran the plates. He was arrested and fingerprinted, which gave away his identity as Ted Bundy, wanted for murder in several states out west.

While Bundy's own revelations were often dubious, it appeared that his adult crime spree had begun in 1973 or 1974, when he killed more than two dozen girls in western states. The first apparent victim might have been Kathy Devine, fifteen, who was running away from home. Then there was Linda Ann Healy, who turned up missing in Seattle in January 1974. Blood drenched her mattress and there was a bloodstained nightgown close to the bed, but no body. Dozens of apparent abduction-murders followed throughout the Pacific Northwest and a suspect emerged who called himself "Ted" and who appeared to lure girls into his car by wearing a cast or sling on his arm and acting charming but helpless.

Several young women vanished altogether, while others were found dumped in remote places, such as densely wooded hillsides. Similar deaths in Utah and Colorado alerted law enforcement to the possibility of a transient killer. With the assistance of an eighteen-year-old woman who was nearly a victim but managed to jump out of Bundy's car, they apprehended him. She said he had posed as a police officer to get her to accompany him. Tried first in Salt Lake City, Utah, in 1976, Bundy was convicted of aggravated kidnapping and sentenced to one to fifteen years in prison. Then Colorado charged him with murder and he asked to manage his own defense. Given access to a law library, Bundy slipped through a window and ran. Recaptured eight days later, he managed once again to escape (by losing enough weight to fit through an opening in his cell), and this time he left the Western states and went to Florida.

At the police station after his arrest in Tallahassee, the investigators requested that Bundy provide a dental impression that they could use to compare to the suspicious bite mark on victim Lisa Levy, but Bundy refused. They got a search warrant that authorized them to get the impression in any way they could, and made a surprise trip so as to prevent Bundy from grinding his teeth down in an effort to disguise his bite. During this examination, Dr. Richard Souviron, a dentist from Coral Gables, took photographs of Bundy's front upper and lower teeth and gums. He noted the uneven pattern, which he knew would facilitate making a match.

Bundy acted as his own attorney for most of his trial, until Dr. Souviron took the stand. At that point, Bundy sent in the lawyer assigned to assist. This same lawyer had already requested that the bite-mark evidence be thrown out because there had been no grounds for the warrant. The judge had ruled the evidence admissible.

In the many different analyses preceding the trial, the tissue from Lisa Levy's buttock had been largely destroyed, but the photograph with the ruler still remained. Souviron described the bite mark on Levy as the jury examined the photographs. He pointed out how unique the indentation mark was and showed how it matched the dental impressions of Bundy's teeth. He demonstrated the structure of alignment, the chips, the size of the teeth, and the sharpness factors of the bicuspids, lateral, and incisor teeth. Then he put an enlarged photo of the bite mark from Levy on a board and laid over it a transparent sheet with an enlarged picture of Bundy's teeth. There appeared to be no doubt that Bundy had bitten her that night in some mad frenzy.

Souviron went on to explain that there had been a double bite: The attacker had bit once, then turned sideways and bit again. The top teeth remained in the same position, but the lower teeth left two rings. That gave Souviron twice as much to work with to prove his case. When questioned by the defense about the subjective nature of odontology interpretation, Souviron explained that he had done several experiments with model teeth to be assured of the standardization of his analysis. The attorney pointed out that the ruler in the photo had been lost, but Souviron countered with the obvious fact that it once had existed because it was in the photo.

Then the state called Dr. Lowell Levine, the chief consultant in forensic dentistry to New York City's medical examiner. He testified that for the marks to have been left on the skin in the manner evident, the victim had to be lying passive, probably already knocked unconscious or killed. He also educated the jury on the long history of forensic odontology.

Along with the eyewitness testimony of Nita Neary, the bite-mark evidence was the best the prosecutor had. On July 23, Bundy was found guilty and sentenced to die in the electric chair. This was the first case in Florida's legal history that relied on bite-mark testimony, as well as the first time that a physical piece of evidence actually linked Bundy with one of his crimes. In February 1980, he was tried for the murder of Kimberly Leach, whose remains had been found two months after she was abducted, and he was convicted for this one as well, receiving a third death sentence. Afterward, he confessed to the murders of thirty women over a span of four years, although experts estimate there were more victims. He was executed in 1989.

ASSISTANCE FROM THE DEAD

Among the difficulties of cases like this are victims found so long after their murders that time of death estimates are difficult to make. Dr. William Bass III, a forensic anthropologist for the State of Tennessee, discovered this himself one cold day in December 1977, which precipitated a rather unique and innovative response to the problem. A detective asked him to estimate the age of a set of human remains found on some family property. It was lying in the grave of a former Civil War colonel, William Shy, who had been killed and then buried in 1864. The police suspected that someone had used the grave as a dumping ground for a more recent death. Bass arrived with the police and could smell for himself the presence of the dead. Not three feet below the surface, where deputies had dug a hole, lay a man without a head and dressed in a tuxedo.

Bass agreed to go into the grave, which reeked of decay, and carefully remove the remains. It took him four hours in the freezing cold to maneuver them from the tight squatting position he was in. He handed the parts out as they came undone, and then laid them on a plank. The flesh was pinkish, and he estimated that the remains were those of a white male between the ages of twenty-four and twenty-eight, and since the flesh still retained a powerful odor and was fairly intact, this person had been dead between six months and a year. Bass said that he would need to examine the remains more closely in his lab. Then he went back into the grave to retrieve the head by having the officers hold him by the feet so he could rummage around upside down inside the broken cast-iron coffin. What he saw with a flashlight was evidence of the remains of Colonel Shy, so he left them alone and returned to the surface without the head. Wherever it was, the search could wait. Night was falling and it was getting quite cold.

Bass transported the remains he had back to his anthropology lab at the university. There he attempted to boil the flesh from the bones, but it did not smell like what he was used to from bodies discovered in the wilderness. The clothing, too, did not fit the scheme of a man recently deceased and Bass soon realized from the age of the material and its fashion that he had made a rather dramatic mistake: The corpse was that of Colonel Shy himself and he had been dead and buried some 113 years. Bass had been correct about the age estimate—Shy had died at the age of twenty-six—but not about the time since death. Because Shy's body had been embalmed with arsenic, boiling it had produced the odd odor. He then heard what the police had found inside the coffin: a jawbone, several bone fragments, and a full set of teeth. Assembled, this collection of gruesome items proved to be the missing head. The man's remains had seemed recent because he'd been well-preserved in the cast-iron coffin. Apparently, vandals seeking Civil Wear artifacts had disturbed the grave, so the remains, exposed to the air, had begun to decompose more rapidly.

Bass, who had begun his professional work for the Smithsonian Institution in Washington, D.C., cataloging the bones of Native Americans, knew that something had to be done to provide a scientific basis for estimating time since death from human remains. He requested a three-acre piece of land from the University of Tennessee at Knoxville so that he could study decomposing human bodies. He knew that if professionals in this field were to learn about decomposition rates, they'd have to find a way to study them quite rigorously in various actual conditions under measured controls. Thus, in 1980 near the university, he broke ground for the project, which he called the Anthropology Research Facility. Law enforcement eventually renamed it the Body Farm.

In 1981, Bass laid out the first body, an unclaimed cadaver. He meticulously documented the conditions for its decomposition, and as he acquired more specimens, he placed them in other contexts: submerged in water, buried in earth, left inside buildings, locked in the trunks of cars. As they decomposed, they provided information about what happens to corpses under different conditions. From insect analysis to the nuances in odor at different points during the death process to death-related bacteriology, there seemed no end to the types of experiments that could be done to assist law enforcement. The researchers expanded in number and specialization, and the Body Farm became a center for training and consultation in difficult cases, including for the FBI.

But the most dramatic forensic discovery was just on the horizon, growing out of the work done in molecular biology over the past century.

TWELVE

SHAKE-UP IN IDENT

CODE SECRETS

In 1980, Dr. Ray White, a biologist working at the Howard Hughes Medical Institute in Utah, devised an easier way to find genes with a technique that was called restricted fragment length polymorphism (RFLP). He was able to use a restriction enzyme to cut repeated patterns of DNA and determine how often they showed up in the same pattern. In certain spots, this sequence varied from person to person, and the segments were used as markers for specific genes. Soon, this technology would have a new application, similar to another one about to be developed.

In Mendocino, California, Kary Mullis was troubled by his work with sickle cell anemia and the difficulty he'd experienced in using the small amounts of DNA typically available. In a flash of inspiration, he envisioned the ability to select specific segments of the DNA molecule and clone or replicate them millions of times until there was enough to work with. Tests could then be applied to the copies. The tools were already available for accomplishing it, so he set out to bring the idea into practical reality, calling it polymerase chain reaction, or PCR. Scientists quickly applied it in the field of medicine, but no one just then thought about its possibilities for crime investigation.

During this time in the field of document examination, a forgery succeeded in embarrassing a number of handwriting professionals, yet when the fraud was exposed, the incident ultimately proved a triumph for other types of examiners in this field. A German publishing company, Gruner & Jahr, was offered a collection of sixty handwritten notebooks—reportedly the lost diaries of Adolf Hitler, removed from Berlin after World War II on board an airplane, which had crashed. Nazi document collector Konrad Kujau said that farmers had discovered them, and the notebooks went through several hands until he acquired them via a general in East Germany. Kujau had brought them to journalist Gerd Heidemann, who was on the staff of *Stern*, a newspaper owned by Gruner & Jahr. He acted as Kujau's agent and the publisher agreed to pay $2.3 million for the lot, including a heretofore undiscovered third volume of Hitler's two-volume book, *Mein Kampf.* Experts read them and concluded that Hitler had been oblivious to "the final solution," used to exterminate millions of people. It seemed that the history books would have to be significantly revised.

Stern began serializing the documents, selling publication rights to *Newsweek* in America and to the London *Times*. It was the owner of the *Times*, after having serious doubts, who insisted that tests be performed to establish the authenticity of the diaries, but even after the initial round of testing, the answer was still unclear.

Three experts accepted the task of comparing the black notebooks with their special seals against samples of handwriting affirmed as Hitler's: Max Frei-Sultzer, the former head of the forensic science department for the police in Zurich, Switzerland; Ordway Hilton, a specialist in document verification; and the third man worked with the German police. All agreed that the same person had written all of the texts, and this author's handwriting was consistent with that in the comparison samples. They stated that if the exemplars were indeed Hitler's, then the diaries appeared to be authentic.

However, the handwriting analysis was not the only field of expertise applied to the documents. Forensics tests on the paper and ink proved something else altogether. Paper is classified via its composite materials, differing according to specific additives, the presence or absence of watermarks, and its surface treatments. Specialists can determine the date when a particular type of paper was introduced, based on what was known about paper production from different eras and cultures. They can also analyze ink for its specific components with microspectrophotometry or thin-layer chromatography, to determine whether it's made from iron salts, carbon particles, or synthetic dyes, and what the various additives might be. From an extensive database kept at the U. S.

Bureau of Alcohol, Tobacco, and Firearms (BATF), examiners can provide a probable origination.

The West German police put the paper under ultraviolet light and subjected it to other tests, finding a component additive in use only since 1954—after Hitler had died. The threads attaching the seals contained material manufactured after the war, and the ink used had not been available at the time the diaries were purportedly written. One test that involved the evaporation of chloride proved that the documents were quite recent.

In addition, content analysis undertaken by historians revealed glaring mistakes, apparently overlooked by the original analysts in the attempt to keep the diaries secret until publication. One distinguished historian of Hitler's regime, Hugh Trevor-Roper, had even vouched for their authenticity. But clearly he'd been mistaken.

At no point during all of this processing had anyone thought to research Kujau's background, and if they had, the publisher might not have been so eager to purchase the documents. As a child, Kujau had sold forged autographs of famous politicians for pocket change. Later he'd manufactured so-called Nazi mementos, including an introduction to a sequel to *Mein Kampf*. He had not named the person from whom he'd purchased the documents, and as it turned out, he had forged the samples the experts had used as comparison exemplars of Hitler's handwriting—obviously quite successfully. This revelation proved both outrageous and humiliating.

When the forgery of the Hitler diaries was exposed in

1984, Kujau tried to flee, but he was arrested and tried in Hamburg. He confessed and served three years in prison.

The duped experts learned that during the course of the two years in which Kujau had worked on the diaries, he'd "authenticate" them by hitting them with a hammer and staining them with tea leaves. His original intent had been to write only one book, but when his agent managed to seal such a handsome deal with this publisher, Kujau wrote more, relying for content on newspapers, medical reports, and reference books that contained transcripts of Hitler's speeches. He'd even forged a letter from Hitler, in case he needed it, giving him authority to compile the diaries for posterity. But in writing the more recent documents in the effort to further enrich himself, Kujau had reached too far, making himself vulnerable to those scientists who insisted on careful analysis and impeccable provenance.

Also in 1984, a case in England coordinated several forensic areas. Graham Backhouse had complained about harassment for several months, receiving threatening notes in the mail. Then it grew more serious. One morning he asked his wife to go pick up antibiotic for the sheep, so with her car out of commission, she got into his Volvo. As she started the engine, the car exploded. She survived but lost part of her thigh. Then another threatening letter arrived, which thwarted the document examiner, because it had been written and then retraced. The earlier note was easier to read, but there were no exemplars as yet with which to compare it.

The police tried to protect Backhouse, but on April 30, his alarm went off. When they arrived at his farm, he was standing over a body at the foot of the stairs. Covered in blood, he claimed he'd shot his neighbor, Colyn Bedale-Taylor, with whom he'd been feuding, in self-defense when the man tried to attack him with a knife. Indeed, he had a gash across his shoulder and knife wounds on his face to prove it. The police found part of the pipe that had been used for the car bomb on Bedale-Taylor's property, which seemed to seal the case.

Yet when forensic biologist Geoff Robinson examined the bloodstains where the alleged attack had occurred, he found them to be the wrong shape to support Backhouse's story. They were round, as if they had dripped straight down rather than flung or cast off from a person struggling or in flight. Doubts about Backhouse's tale began to surface. In addition, it appeared that he had not tried to defend himself when attacked and had left no trail of blood, though he claimed he'd been wounded while running for his shotgun.

Dr. William Kennard, the pathologist, thought it was odd that the knife used in the attack was still in the hands of the deceased. That seemed impossible. Also, some of the blood on Bedale-Taylor's shirt was matched to Backhouse, and it had dripped straight down, as if Backhouse had been standing over the body while bleeding.

Robinson then tested the envelope in which one of the letters had come. He found wool fibers consistent with a sweater that Backhouse wore and on a notepad in his house, a document examiner made out the impressions of a doodle found on the other side of one of the notes. Clearly, Backhouse himself had written the harassing notes, planted the bomb (putting his own wife at risk), and invited his victim to his home to kill him and stage the scene. The entire affair was a calculated setup. With the help of experts in biology, explosives, pathology, and document examination, this case turned on its head and the real killer was apprehended. Backhouse got two life terms for the murder and the attempted murder of his wife (for insurance money, it turned out).

THE WORLD HAS CHANGED

Even as the scandal over the diaries cooled down, molecular biology as a forensic tool was coming to life. In 1984, Dr. Alec Jeffreys, a British molecular biologist, used RFLP as the DNA-typing protocol to dissolve an immigration dispute over a boy from Ghana who claimed he had a British mother and wanted to live with her. Jeffreys also resolved another parental issue with paternity testing that affirmed that a French adolescent was the father of a British-born child. For this work, Jeffreys received many public honors and in an article in *Nature* in 1985, he named the process he used "genetic fingerprinting," stating that an individual's DNA pattern was unique and would not be found in any past, present, or future person. This put him in demand for more such cases. But the application that would impress the world and change legal history most dramatically occurred in England the following year, although it involved a criminal investigation that had been in process since 1983. (There was in fact a rape case that was resolved with DNA before this one, but it did not generate the same degree of international response.)

On November 23, in the village of Narborough, fifteen-year-old Lynda Mann had traipsed along a path known as the Black Pad, and there she met a man intent on committing a sexual offense. He also killed her. Mann's body was found near the path the following day, raped and strangled. The only clue the police could collect was a semen sample, which revealed the killer's blood as type A. Soon the case went cold, but it remained important to the local police, especially when Dawn Ashworth, also fifteen, was found raped and murdered in 1986 on another footpath only a mile from where Lynda Mann had met her demise. The attack had been shockingly violent and aggressive, and the semen from her remains proved that the blood type matched that from the earlier incident, which increased the chances that the same man had killed both girls.

Then a seventeen-year-old kitchen porter was arrested and after a lengthy interrogation of some fifteen hours, he confessed to the second murder but denied involvement in the first. After he provided details for the Ashworth incident that had not been published, admitting that perhaps he had just "gone mad," investigators felt sure he was good for both. They decided to seek out Jeffreys, whose lab was only six miles from the first crime scene, to ask him to apply his genetic fingerprinting to confirm the boy's involvement.

Jeffreys had read about the two murders in the newspaper, so he eagerly agreed to test the semen samples. Since the one from the Mann murder was fairly degraded, he was uncertain what to expect. But he put it through the lengthy RFLP process and awaited the results.

In RFLP testing at that time, after the extracted DNA was cut into fragments, the fragments were covered in a gel to separate them into single strands. An electrical current was applied to push the negatively charged fragments through the gel at speeds relative to their length toward the positive pole, with the shorter pieces migrating faster. There they lined up according to size. The pieces were removed from the gel with a nylon membrane called a Southern Blot, and the DNA fragments were fixed to the membrane. This process exposed the A, T, C, and G bases, which could then be treated with a radioactive genetic probe. The single-strand probe would bind to its complementary base, revealing the DNA pattern, and a multilocus probe would bind to multiple points on multiple chromosomes. The probe identified specific areas of the DNA with dark bands, as revealed by an x-ray (autoradiograph or autorad) of the membrane. Then a print of the polymorphic sequences made it possible to compare to prints similarly gained from other specimens. The interpretation of a sample was based in statistical probability. If four fragments were identified, then the probability of each occurring in the population was multiplied by that of the other samples.

In the radioactive membrane, the genetic profile of Lynda Mann's rapist was revealed, but when this result was compared to the porter's sample, there was no match. However, the work continued for the next "nail-biting" week on the semen sample removed from Dawn Ashworth, and this was compared to that from Lynda Mann. This time there *was* a match, but not the one expected. The samples matched each other, so the same person had committed both crimes. However, neither sample implicated the young suspect. Despite his confession, he was not their man.

The police who had worked long hours on the case wanted to challenge this finding because it made no sense to them, but they couldn't. None even understood the process, and Jeffreys was one of the few people in the world who knew what he was talking about. The officers could only admit that they were wrong; thus, their chief suspect became the first person in criminal history to be freed based on a DNA test. Yet there was still the matter of his confession. When asked why he had admitted to rape and murder, he said that he'd felt pressured. Indeed, he'd confessed to a number of other inappropriate incidents as well. Investigators believed that he had probably come upon the body that night, which explained how he knew unpublished details about the crime scene, and they had simply worked him into a confession based on that. (Others have suggested that during the interrogation they inadvertently fed him the details.)

While the investigation continued, Jeffreys traveled to the FBI's academy at Quantico to show them what was involved with the process, and at the age of thirty-six he became an international celebrity. Many people became interested in how they could cash in. But Jeffreys' fame at this point was minor in comparison to what it was soon to become.

Back in Narborough, the police were determined to find the right perpetrator, so the men of Narborough and villages nearby within a certain age range were asked to voluntarily provide a blood sample. More than 4,500 men agreed to do it, most of whom were eliminated via conventional blood tests (since DNA analysis was expensive and time-consuming). But the real goal was to ferret out any man who would not willingly submit, because he might have something to hide. Yet after all this processing, the police were disappointed to realize that they had failed to identify the Footpath Murderer. Then in September 1987, they learned about a suspicious incident that had occurred over a month earlier.

A baker named Ian Kelly claimed that he'd provided his own sample to the police as a substitute for his friend and fellow baker, Colin Pitchfork. He'd received a falsified passport and had gotten away with it, so he bragged about it during lunch at a pub. A female manager overheard Kelly and passed this information along to the police. They knew that Pitchfork had been arrested numerous times for indecent exposure, so given this background and his attempt to circumvent their investigation, on September 19, they brought him in for interrogation. He admitted that he'd been out looking for girls to whom he could expose himself when he happened across the victims. There had been no witnesses, so he'd seized the opportunity to rape them. To ensure that he could not recant, the police sent Pitchfork's blood for DNA testing, and the results proved that Pitchfork's genetic profile was indistinguishable from that of both semen samples. He became the first person to be convicted of murder based on genetic fingerprinting. Pitchfork drew a life sentence, while Kelly received a suspended sentence for obstructing the investigation.

The Pitchfork case sparked headlines around the world, as well as a book by bestselling writer Joseph Wambaugh, *The Blooding*, inspiring a great deal of attention from the law enforcement community. It seemed that a potentially foolproof method was at hand for solving crimes involving biological evidence. The rush was on in many different places to apply DNA technology to more crimes. Lifecodes, located near Westchester, New York, became the first private lab in the United States to offer RFLP testing for criminal incidents.

Around this same time, crime investigation took several large steps in other directions. The FBI had developed a national computer database called the Violent Criminal Apprehension Program (VICAP), slating it to become the most comprehensive computerized database for linking and solving homicides nationwide. Police departments around the country were invited to record solved, unsolved, and attempted homicides; unidentified bodies in which the manner of death was suspect; and missing-persons cases involving suspected foul play. By

1985, the Bureau had also set up the National Center for the Analysis of Violent Crime (NCAVC).

State and local agencies built up automated fingerprint identification systems (AFIS) via computers, and the FBI expedited an exchange of information among law enforcement agencies by introducing a standard system of fingerprint classification that harmonized the transmission of information from one AFIS system to another. The program digitally encoded scanned prints into a mathematical algorithm based on their characteristics and the relationships among their features. Each image received a corresponding file of demographic data. In contrast to the days and weeks it once took a fingerprint analyst to complete, within seconds the computer could compare a set of prints against a half million others.

Nevertheless, for cases with biological evidence, the process of identification was about to evolve. After the initial success with DNA typing in England, Lifecodes quickly developed the technology in the States. In 1987, Florida's assistant state attorney, Tim Berry, contacted Lifecodes's forensic director, Michael Baird, about a rape case that was going to court. He wanted to know what DNA identification could do for him.

The case had begun in May 1986, when a man entered the Orlando, Florida, apartment of Nancy Hodge and raped her at knifepoint. She managed to see his face before he left, grabbing her purse on his way out. During the succeeding months, the man raped more women, taking care to conceal his identity, and on his way out he always took something that belonged to them. In the course of six months, investigators believed this man had raped more than twenty-three women, but he proved to be maddeningly elusive. Eventually, however, he made a mistake: He left behind two fingerprints on a window screen. When another woman reported him as a prowler, the police chased him for two miles. Once they had him in custody, they found that his prints matched those from the window screen. Finally, they had their man: Tommie Lee Andrews. In addition to the fingerprint evidence, Nancy Hodge, a victim who had seen him, also identified him.

Yet proving Andrews to be a serial rapist with a high number of victims was going to be much more difficult. Although Andrews's blood group matched semen samples taken from several of the victims, blood group analysis was not precise and thus there was potential in each case for reasonable doubt. Berry hoped a DNA match might strengthen the prosecution's position, as well as close numerous cases and send Andrews to prison for a much longer stint.

Blood samples from Andrews and semen samples from the rapist went to Lifecodes. Within two months, the results came back: The bar codes from the rape samples were too highly consistent for the semen to have originated with anyone other than Andrews. The odds were stacked against it.

Nevertheless, DNA testing had not yet been accepted into court and before it could be used, it had to go through a pretrial hearing. Ever since the *Frye* test in 1923, any new scientific technology introduced as testimony had to pass the test of acceptability within a relevant scientific community. That way the courts avoided admitting evidence based on whim or supposed science that was actually devoid of objectivity and rigorous controls. DNA analysis had to prove itself scientifically sound in method, theory, and interpretation, and be positively reviewed by peers.

The hearing was long and complex, but finally the judge admitted the DNA evidence. However, Berry made a misstep by stating impressive odds for the samples to have come from Andrews—one in ten billion—and he could not substantiate how these odds had been derived. He withdrew the figure, hoping to repair the damage, but the jury hung. Andrews went to trial for the second rape charge and this time he was convicted. Months later, the first rape charge was retried and the DNA evidence was brought in with more clarity and power. After that trial, Andrews's prison sentence stretched from his initial twenty-two years for rape to 115 years for serial rape. He became the first person in United States history to be tried and convicted with DNA evidence.

From there, DNA gained increasing acceptance in the courts, although challenges were aimed at the way samples were interpreted, as well as at shoddy handling of specimen evidence. Without safeguards in place for proper scientific examination, the labs put prosecutors at a disadvantage, because defense attorneys were learning about the vulnerabilities in the system. Manhattan-based attorneys Barry Scheck and Peter Neufeld cofounded the DNA

Task Force of the National Association of Criminal Defense Attorneys. Their goal was to debunk DNA typing in courts across the country, and failing that, to at least limit its application. They evaluated laboratories and evidence technicians, offering support to any attorney faced with this evidence in the courtroom. Because many different things can occur between the collection of a sample and the final interpretation, the courts were forced to review DNA testimony on a case-by-case basis. The FBI reported the first RFLP-processed case from its own lab in 1989, publishing guidelines to set standards for quality assurance.

That year, in *People* v. *Castro*, the technology was successfully challenged for the first time. On February 5, 1987, Vilma Ponce and her two-year-old daughter, Natasha, were stabbed to death in their home in a Bronx apartment building and a speck of blood was found on the watch of a neighbor named Joseph Castro, who was the building's handyman. Ponce's common-law husband, David Rivera, identified him as the man he'd seen leaving the building on the day of the murder, covered in blood. Right afterward, Rivera had discovered the door to his apartment unlocked and open, and the police arrived to find his pregnant wife stabbed fifty-eight times, while the child had received sixteen wounds.

Castro was the obvious suspect, especially after an acquaintance of Ponce's reported that Castro had pestered Ponce with sexual advances, which she had spurned. He was also the worker who had installed the lock on her door, which proved to be defective. However, information turned up that Rivera had been violent with Ponce in the past, breaking her jaw, which complicated the case. Bronx prosecutor Rise Sugarman hoped that the blood from Castro's watch would prove to be from one of the victims, because it would add weight against Castro. Lifecodes tested it, along with samples from the victims, and reported that it matched Vilma's at three locations on three chromosomes.

However, during the course of a twelve-week hearing, the defense, aided by Neufeld and Scheck, pointed out that the lab had made a technical error, which invalidated the DNA results. The defense's expert was Dr. Eric Lander, an MIT scientist, who helped the attorneys get a concession from Lifecodes that they did not in fact utilize mathematical standards for reporting the odds but instead had interpreted the results subjectively, via observation.

As the issues grew more complicated, scientists from both camps agreed to meet outside the legal arena to discuss the problems. Together they devised a two-page statement that admitted that the results in the Castro case were too poor to make a definitive analysis either way, and that most scientists would agree with them: The results should be considered inconclusive. One of the prosecutor's experts even returned to the stand to qualify his original testimony.

The court decided that the testing could be used to show that the blood on the watch was not Castro's (to exclude him), but it could *not* be used to claim that the blood matched one of the victims (to include her as the source). Given the room this decision left for reasonable doubt, the DA offered a deal. Castro pled guilty to both murders for a lesser sentence, which to some extent vindicated the Lifecodes analysis.

Yet thanks to cases like this, whose individual failings some viewed more generically as the failings of DNA technology, and to alleged statements in the *New York Times* in 1990 by prominent scientists against DNA testing, the courts backed up. Although four of the scientists named insisted they had made no such statements (their written objections were ignored), the courts grew more conservative, allowing the use of DNA to *exclude* suspects as the source of origin but not to make claims that the suspect's DNA was a match to collected evidence. Prosecution experts had to work hard to prove that DNA analysis could perform as promised.

Improved methods over the next few months increased the testing accuracy, and the technology that could demonstrate the chance that a specific sample matched a specific person showed statistical odds so overwhelming that the courts gradually allowed it for stronger claims. In 1992, the National Research Council Committee of the National Academy of Sciences affirmed the use of DNA analysis but recommended tight regulations over collection and processing. In addition, they advised that the testing be done by objective parties with no stake in the outcome.

That same year, Connecticut prosecuted a murder case with DNA evidence alone—a first. Carla Almeida, a twenty-two-year-old masseuse, had turned up missing on April 18, 1988, and a client with whom she'd been booked, Tevfik Sivri, said that she'd come and gone. Criminalist Henry Lee went to Sivri's home and found nothing out of the ordinary, except that the place appeared a little too clean as if in an effort to remove evidence. He searched the house extensively, then reached down to feel the carpet. It was damp. Using a blood reagent, he showed that the carpet had been soaked in blood at some point. When it was cut away, a large pool of blood was revealed on the bare floor. Lee used a calculation of his own devising to prove that the loss of this much blood meant the loss of life. The police then found a spot of Carla's blood in Sivri's car, proven with DNA analysis. Although the state had no body, they went to trial against Sivri. The judge approved the analysis as reliable, and the jury convicted Sivri.

However, in 1994, he got a new trial. But by that time, Carla's remains had turned up. She'd been shot in the head and dumped near a tree farm. Sivri was again sentenced to life.

REVELATIONS

In July 1984, items of clothing from a missing nine-year-old girl, Dawn Hamilton, were found in a tree in the woods outside Fontana Village, Maryland. Then the police found the girl's body lying facedown not far from the path, still warm to the touch, but clearly dead. She had been raped, strangled, violated with a stick, and bludgeoned with a rock. An anonymous call to the police sent them to investigate Kirk Bloodsworth, because the caller reported seeing Bloodsworth with the victim the day before she was murdered. Despite his insistence that he was innocent and did not know the girl, Bloodsworth was arrested. Other witnesses offered police enough detail for a sketch, and from it yet another witness identified Bloodsworth. All five appeared at his trial to state that he was the man they had seen with the victim, and a shoeprint found near the victim proved to be his size. For further circumstantial evidence, prosecutors said that Bloodsworth had told a friend that he had done something terrible that day that would adversely affect his relationship with his wife, and during the interrogation, he had mentioned a bloody rock. The case was stacked against him so it was no surprise when the jury convicted him. They also sentenced him to death in the gas chamber.

On appeal, Bloodsworth's attorney said that the police had shown him the bloody rock during interrogation, and the incident to which he had referred regarding trouble with his wife was that he had failed to purchase food as she had requested. In addition, there had been another suspect, but the police had failed to inform the defense about this development. The conviction was overturned, which gained Bloodsworth a new trial.

But the same witnesses testified in this second trial and in 1988 the same result ensued: Bloodsworth was once again convicted, but this time received two life sentences, to run consecutively. Bloodsworth assured everyone that he was innocent and read everything he could find about legal procedure, seeking something that might provide a way to clear his name and free him. He came across Wambaugh's book, *The Blooding,* about Colin Pitchfork and genetic fingerprinting, noting that the first suspect, who had given a confession, had been exonerated. Bloodsworth called an attorney who then contacted attorney Bob Morin, who specialized in death penalty cases, and Bloodsworth asked Morin to try to get DNA testing for him. Centurion Ministries helped him to acquire court approval for the test, granted in 1992. Morin contacted Cellmark, the second private lab in the country to offer the method, but their RFLP method required more material for analysis than was available. Morin then looked to Kary Mullis, the inventor of PCR, but he had left Cetus Corporation, which owned the rights and had exclusively authorized Dr. Edward Blake of Forensic Science Associates (FSA) to utilize it. Morin sent the victim's shorts and underwear, the stick used on her, and an autopsy slide of a semen sample to Blake, who after three months indicated that there was semen on the panties, despite the FBI's report that there was not. Using PCR-based DNA testing, FSA determined that the amount of spermatozoa on the slide had proven insufficient for testing, but analysis of the stain on the panties excluded Bloodsworth.

Morin and Bloodsworth were thrilled, but the ordeal wasn't yet over. The FBI wanted to run its own tests. Morin got Barry Scheck involved, who confirmed Blake's findings, so he resumed his confidence, but it meant more time in prison for Bloodsworth. Finally, the FBI let him know what their tests indicated: Bloodsworth was excluded as the source of the semen.

On June 28, 1993, after nearly nine years served in a dangerous hellhole for an act he did not commit, Kirk Noble Bloodsworth was released from prison. Later that year he was granted a pardon and the State of Maryland paid him $300,000, based on ten years of lost income. Despite the collective eyewitness testimony, offered twice, Bloodsworth became the first person to be exonerated with DNA technology from death row. And there would be more, eventually shocking the governer of Illinois so much that he placed a moratorium on the death penalty in his state for further review. Scheck and Neufeld set up the Innocence Project in the Benjamin N. Cardozo School of Law at Yeshiva University in Manhattan to assist falsely imprisoned people to acquire post-conviction DNA testing.

While Bloodsworth was walking free, another man who probably believed he'd never be caught became part of the first case in which the practice of DNA data banking solved a crime. The victim was Jean Ann Broderick, and she had recently moved into the neighborhood of Lowry Hill in Minneapolis, Minnesota. One night in November 1991, she and her roommate walked home, passing a halfway house for sex offenders, most of whom had been required to provide blood samples for the state's new DNA data bank. The next day, Broderick was found in her room, raped and strangled. There were no leads, aside from semen left behind on her body. The authorities went through the data bank and found a DNA profile for an illegal immigrant, Martin Perez, which matched the semen. He had a record for rape, burglary, and assault in several states, and in fact should have been in jail at the time of the murder for a burglary, but he had used fake credentials to elude conviction.

A student who had lived in the apartment just prior to Broderick identified Perez as a burglar who had broken into the place that summer. Perez was arrested and tried, and it was the DNA from the data bank that proved to be the most compelling evidence against him. Had there been no such collection of samples, Perez would not have been apprehended for this crime. The identification via DNA had allowed the police to develop a case with other types of evidence. By the end of 1993, twenty-one states had passed laws that required sex offenders to provide blood samples for similar data banks.

There was another first for DNA that same year in Phoenix, Arizona. Denise Johnson was found on May 2, 1992, strangled and left nude near a cluster of palo verde trees in a remote part of Maricopa County. She had been bound with a cloth tied around her neck. A pager belonging to Earl Bogan was found nearby and a witness had seen a specific type of white truck leave the area—the same type driven by Earl Bogan's son, Mark. The truck was seized for a search, which turned up seed pods in the back from a palo verde tree, but nothing that indicated that Johnson had been inside. Nevertheless, Mark Bogan admitted that he'd picked her up for consensual sex, but after they'd argued, he had dropped her off. He denied being anywhere near the area where she had been found, but when confronted with the pager, he said that she had stolen it from him. It seemed a tough story to break, but the seeds provided a possible avenue. Investigators looked for a way to use this potential lead.

Dr. Timothy Helentjaris, a professor of molecular genetics agreed to test the pods from the truck. Using Randomly Amplified Polymorphic DNA (RAPD), a technique known for several years among plant geneticists, he compared the various trees in the area against one another and managed to match the crime-related pods to a specific tree. This finding tied Brogan's truck to the place where this plant grew—exactly where the body was found. That same tree showed a recent gash in its trunk, which was precisely where the truck's bumper would have hit had Bogan backed into the tree. With all this evidence against him, Mark Bogan was convicted of first-degree murder. An appeal challenged the uniqueness of the RAPD method, but it failed and the conviction was upheld.

Then a court case that reexamined the nature of scientific evidence affected the entire field of forensic science. Jason Daubert and Eric Schuller were born with serious birth defects. Their parents alleged in a suit against the pharmaceutical company that the mothers of both children had ingested Bendectin to fight nausea while they were pregnant. The suit ended up in federal court, where the pharmaceutical company attorneys insisted that the drug did not cause birth defects in humans. Dr. Steven Lamm offered testimony as their expert that upon reviewing more than thirty published studies involving more than 130,000 patients, he found no evidence that the drug caused malformations in fetuses.

But the petitioners offered eight experts of their own who concluded that the drug can indeed cause birth defects. They had used test tube and live animal studies, as well as chemical studies of drugs that bore a structural similarity to Bendectin and that did cause birth defects. They also reinterpreted earlier studies that had found no link to give those studies a different spin. The court was faced with setting forth conditions under which scientific evidence is admissible. They decided that "scientific" means having a grounding in the methods and procedures of science that are sufficiently established as to have general acceptance in the field, and any claim of having "knowledge" must be stronger than subjective belief. The petitioners' evidence was deemed to have fallen short of this mark. Their experts did not sufficiently show causation between the drug and the defects with proven methods, and their approach had not been subjected to peer review. In addition, their methodology diverged significantly from that which was generally accepted in the scientific community. Thus, it could not be considered reliable.

In its 1993 decision, the U.S. Supreme Court gave a nod to the *Frye* test from 1923, stating that while the notion of "general acceptance" had some problems, most courts still relied on it. They also noted the Federal Rules of Evidence, which superseded the *Frye* test, did not make an issue of general acceptance. In federal courts, then, the *Frye* test would no longer apply to novel scientific evidence. It was now up to the judge to evaluate scientific reliability. When faced with a decision to admit evidence or not, a judge had only to focus on the methodology, not on the conclusion, and also on whether the scientific evidence applied to the facts of the case. In other words, judges now had to determine whether the theory could be tested in accordance with scientific criteria, the potential error rate was known, the method had been reviewed by peers and had attracted widespread acceptance within a relevant scientific community, and the testimony was relevant to the issue in dispute.

The *Frye* standard was replaced as well in many states by the *Daubert* standard, as cited in *Daubert* v. *Merrell Dow Pharmaceuticals, Inc.* Two cases later in the decade, *General Electric Co.* v. *Joiner* in 1997 and *Kumho Tire Co.* v. *Carmichael* in 1999, would clarify the appellate process from a *Daubert* decision and would apply the same criteria to specialized knowledge involved in other types of technical evidence and expertise.

EVIDENCE ANALYSIS AND CAMERAS IN THE COURTROOM

By the 1990s, forensic science had become a high-tech arena, with more exacting ways to analyze evidence in many areas. With fiber analysis, for example, examiners could use a high-powered microscope to measure the precise diameter and color of a fiber from a crime scene, shine a beam of infrared light to get the absorption spectrum, use polarized light to find refractive indices, or turn to various forms of chromatography to separate dye compounds into specific chemicals. In 1991, Walsh Automation developed an automated ballistics system, and the following year the FBI commissioned the *Drugfire* computerized database to store details about the markings on spent bullets and cartridge cases. Computers were networked to statewide and national databases (even international), similar to an AFIS system for fingerprints. The BATF promoted *Bulletproof* for bullet images and *Brasscatcher* for cartridge cases, while *IBIS,* by Forensic Technology, also offered automated comparisons of evidence images.

However, even as technology improved, the way it was utilized became an issue in a lengthy and highly publicized trial in 1995. DNA was back in the spotlight, as was serology, evidence handling, and crime scene protocol. Called the "trial of the century," it certainly captured the largest audience of any trial to date.

A whimpering dog alerted Sukru Boztepe to the first signal that something was amiss in Brentwood, a wealthy neighborhood of Los Angeles, California. He followed it toward a scene of horrendous bloodshed, so he urged his wife to phone 911. Two blood-covered bodies lay outside the front door of the condominium at 875 South Bundy Drive, occupied by Nicole Brown Simpson, the former wife of actor and former football celebrity O. J. Simpson. The police had been here before, responding to a 911 call concerning domestic abuse. This time they had to deal with something far worse.

Late in the evening of June 12, 1994, someone had attacked Nicole and slashed her to death. Next to her was the body of a man, twenty-five-year-old Ronald Lyle Goldman, who'd been stabbed multiple times. As the story unfolded, it turned out that he may simply have been in the wrong place at the wrong time, bringing Nicole a pair of eyeglasses that her mother had left behind at the restaurant where he worked as a waiter.

Although Nicole was no longer married to Simpson, the police wanted to contact him right away. Going to his home at 360 North Rockingham, detectives found locked gates and no response, so Detective Mark Fuhrman scaled the walls to get into the yard. As he passed by Simpson's white Ford Bronco, parked in the driveway, he noticed a bloodstain on the door. A trail of blood also led up to the house, but Simpson appeared to be gone. It turned out that he had just flown to Chicago. He was notified immediately of what had happened, so he returned to Los Angeles and agreed to answer questions. Investigators noticed a cut on a finger of his left hand. Simpson seemed disturbed by that and told several conflicting stories about how he had gotten it. Yet when the crime scene indicated that the killer had cut his left hand and trailed blood, the two incidents hardly seemed coincidental. In fact, these drops did not match either of the victims' blood types, but an analysis proved they had factors in common with Simpson's blood, with mathematical odds that only one person in 57 billion could produce an equivalent match. In addition, the bloody size-twelve footprints nearby were made by an expensive shoe, a Bruno Magli—a type Simpson owned, in size twelve.

Next to the bodies was a knit hat that turned out to have hair strands consistent with Simpson's hair and a bloodstained black leather glove that bore traces of fiber from Goldman's jeans. The glove's mate, stained with Simpson's blood, was found on his property, and they were similar to the type of gloves sold in a store where Nicole had purchased gloves for Simpson. In addition, the serology associations were strong. Traces of both victims' blood were found inside Simpson's car and house, along with blood that contained his own DNA. His blood and Goldman's were found together on the car's console, and on the carpet was a faint outline of a bloody shoe impression. Socks inside the home showed traces of Simpson's and Nicole's blood.

Then a limousine driver hired to pick Simpson up that evening for the ride to the airport reported that he'd not seen the Bronco and had been unable to get Simpson on the intercom, but had then spotted a black man cross the driveway and go into the house. At that point, Simpson answered the intercom and said that he'd been asleep.

As prosecutors Marcia Clark and Christopher Darden built their case, based on a history of abuse and the notion of Simpson's extreme need to control Nicole, they located photos of a battered Nicole and diary entries that attested to Simpson's stalking behavior. They also found a photo of Simpson wearing a pair of gloves similar to those found at both locations. The clothing he'd worn on the night of the murders had disappeared, as had a bag that he'd refused to let a friend help him with as he'd loaded it into his car that night.

Simpson was notified that he would be arrested for murder, so he fled in his Bronco with his friend Al Cowlings, hinting in a note left behind that he might kill himself. With him were a passport, fake beard, and more than $8,000 in cash. His attorneys finally persuaded him by phone to turn himself in, and he was placed under arrest. Simpson pled not guilty, offered a huge reward for information about the real perpetrator, and hired a defense team of celebrity lawyers. Barry Scheck and Peter Neufeld arrived from New York to be the DNA experts, and Johnnie Cochran took over the lead, while F. Lee Bailey, Robert Shapiro, and Alan Dershowitz filled in the other slots.

The defense team was going to call for a pretrial hearing on DNA evidence, to challenge it from every angle, but decided instead to drop it. In part, they knew that whatever happened could set a dangerous precedent and in part they realized that prolonging the trial process could annoy the jury, whom they wanted on their side. So they waived the proceeding, which many defense strategists thought was a radical decision, and went on with the trial. Barry Scheck felt confident that they could produce challenges in court before the jury that would accomplish all they wanted and also educate and persuade the jury.

The reliability of this evidence was dubbed the "DNA Wars," and as a safeguard, three different crime labs performed the analysis. All three determined that the DNA in several drops of blood at the crime scene matched

Simpson's. It was a 1 in 170 million match using RFLP, and a 1 in 240 million match using the PCR test (which was really just a jump-start procedure to acquire more material). Initially, PCR testing had been considered less definitive than RFLP because it did not detect as many matches at as many locations. However, PCR was faster and soon the testing was refined such that it equaled the reliability of RFLP. It was more practical as well, and could be utilized on much smaller samples. But there was still a problem, which the "Dream Team" of defense attorneys exploited. While they argued that the evidence had been mishandled at the lab, they also had Dr. John Gerdes, a biologist from Denver, explain the highly sensitive nature of PCR testing to demonstrate how easy it was to contaminate. In fact, Gerdes called the Los Angeles crime lab controls into serious question. Kary Mullis was prepared to take the stand as well to say something similar, but he was never called.

Famed criminologist Dr. Henry Lee testified that there appeared to be something wrong with the way the blood was packaged, leading the defense to propose that multiple samples had been switched. They also claimed that the blood had been severely degraded by being stored in a lab truck while the criminalists were processing the scene, but the prosecution's DNA expert, Harlan Levy, said that the degradation would not have been sufficient to prevent accurate DNA analysis. He also pointed out that control samples were used that would have shown any such contamination, but Scheck suggested that the control samples had also been mishandled by the lab—all five of them.

What hurt the prosecution's case more than anything else were the endless explanations they presented via experts of the complex procedures involved in DNA analysis. The defense kept it simple: They accused Detective Mark Fuhrman, who had been at Simpson's home the night of the murder, of being a racist who had planted evidence. When they caught him in a blatant lie about his attitude toward blacks, they added credibility to their claim. They also said that Detective Philip Vannatter had been part of this conspiracy, since he had taken a vial of Simpson's blood to Simpson's home instead of logging it into the evidence room as protocol dictated. In fact, a blood preservative, EDTA, found on a blood smear at the back gate, indicated that someone might indeed have planted blood that had already been processed in the lab. This innuendo was strengthened by the appearance that 1.5 mm of blood that had been drawn from Simpson was missing. No one had a satisfactory explanation.

The evidence against Simpson seemed damning, but the defense team managed to refocus the jury's attention on the corruption in the Los Angeles Police Department. In addition, in response to the prosecution's challenge for Simpson to put on the gloves, he struggled and showed that they were simply too small for his powerful hands. Then Simpson made a clear statement of his innocence, though he was not on the stand. In closing, Cochran disputed the good reputation of the forensics lab, having proven that at the very least, the evidence had been carelessly handled. His mantra, "If it doesn't fit, you must acquit," reminded the jury of the prosecution's fundamental mistake about the notorious glove. Deliberating less than four hours, the jury accepted that something had certainly been wrong with the evidence handling and they freed Simpson with a not guilty verdict. Some of the jurors stated that the prosecution had not made its case.

Yet around America, the case had been an exciting pastime. CNN had covered the trial daily, all day, for nine months, and viewers had been given the impression via a parade of analysts that they were hearing as much as the jury was, so they could therefore make their own determination about Simpson's guilt or innocence. Many heard for the first time about DNA analysis, blood preservative, evidence corruption, and other items related to the world of law enforcement and legal proceedings. Regardless of how the trial turned out, they believed they could judge for themselves what had occurred on the night of the double homicide. Despite the human tragedy involved, the entertainment value was clear and television ratings went through the roof. Some of the participants who had been nonentities were now celebrities, and some fifty books were offered to the public, many becoming bestsellers.

Yet Simpson was not finished. In 1997, he went through a civil trial, with lower standards for the burden of proof, and was found liable for both killings. More evidence against him was introduced, including the fact that he owned a pair of Bruno Magli shoes—the same brand that had left a bloody footprint at the scene. The verdict came as no surprise and those around the country who believed he had murdered his wife found some satisfaction in the order for him to pay the victims' surviving family members $33.5 million, although he never did. Despite this verdict, he was a free man.

During this proceeding, DNA analysis had sustained some damage, and prosecutors in particular had noted the difficulty of presenting such complicated science to juries. It might as easily hinder as help their cases. Nevertheless, as evidence goes, there was little that was as powerful as a strong DNA match, and something was on the horizon that would pressure them to present it, anyway. Juries would soon come to expect DNA evidence, even demand it where none was present. And that would be the result of the media's power. News television had brought Simpson's trial into millions of homes, educating viewers about legal proceedings and expert witnesses. It would not be long before another type of programming would make lay people believe they now understood what the Simpson investigation had been all about.

PUBLIC INTEREST

In 1998, the FBI launched the National DNA Index System (NDIS) a centralized computer database that could link to DNA databases from other criminal justice agencies around the country. It was the capstone of the Combined DNA Index System (CODIS), which used a multilevel software package designed to facilitate computerized cross checks. All states now had profile databases of specific types of convicted felons, and the FBI's CO-

DIS was a vast DNA profile index to which participating labs could submit samples for electronic comparison. Besides the Convicted Offender Index, CODIS also relied on the Forensic Index, which contained DNA profiles from biological crime scene evidence. During the initial experimental phase over a three-year period, the FBI used the system to link nearly two hundred crime scenes to felons.

A search on a database that used only a crime scene specimen and turned up a match became known as a "cold hit," and such searches solved both recent and past cases. Although VICAP was not operating at its expected efficiency because the forms were complex and few jurisdictions were sending information, many cases once believed lost causes were getting solved. Fingerprints from years earlier and DNA from cases with preserved biological evidence were matched to convicted felons.

As the decade, century, and millennium turned, the FBI had acquired more than 65 million fingerprints for AFIS, and had merged its firearms database with that of the BATF to create the National Integrated Ballistics Network. Having comprehensive collections improved identification, but even with improving technology, not everyone took advantage. Yet law enforcement was about to come under the closest scrutiny it had ever endured, for both the better and the worse.

In October 2000, a television series began on the CBS network on Friday evenings, featuring a crew of crime scene investigators on the night shift in Las Vegas. Thanks largely to the exposure received on evidence analysis in the Simpson trial, as well as coverage of other high-profile murders during the 1990s, the television audience was primed and ready for just such a unique series, with all its gore, high-tech equipment, and intrigue. *CSI: Crime Scene Investigation* had a quiet debut, but by the following year, when people the world over watched in horror on September 11 as terrorists flew planes into the Pentagon and New York's World Trade Center, the series had become a phenomenon. Suddenly crime scene investigation became the sexy new topic about which people wanted to know more, and many colleges and universities created programs to catch budding young minds determined to become profilers, investigators, or forensic scientists. Despite the show's inaccuracies, including making evidence handlers into detectives and showing science as a way to get certainty, it spun off two more series in other cities and made forensic science appear to be law enforcement's ultimate weapon.

We might wonder what Lacassagne, Locard, and Vidocq would have thought had they only known how their efforts to get attention for their budding science would one day make it so firmly entrenched in the courtroom and such a mass-market fixation for the public. In some ways, they would have been gratified that science had come so far and made such enormous contributions, but in others, probably horrified. As science blended with fiction, the glittery world of the imaginary CSIs and technology-savvy scientists often misinformed viewers, who were part of the pool of potential jury members. This state of affairs would no doubt have been as annoying to the early pioneers as being misrepresented by journalists from their own time.

Indeed, it disturbed their successors. By 2005, "the CSI Effect" was being debated among scientists and attorneys alike, alarming many staunch members of the American Academy of Forensic Sciences, now more than six thousand members strong, and giving a forum to amateurs who offered only junk science. The dumbed-down world of television forensics was viewed as a threat to the many gains that forensic science had made across two centuries, but on the other hand, the phenomenon inspired innovations as well. Let's give the men who started us along this road their due and examine with eager anticipation what the future of forensic science may hold.

THIRTEEN

THE FUTURE

INNOVATION

The murder had just occurred moments ago, and the police responded immediately. They figured the perpetrator was cornered, since there was no way out of the building. They spread out to protect all exits and wait for him to emerge. But no one came out. Reinforcements arrived and several officers entered the building. They looked everywhere, turned everything over, opened all the closets, and searched all the rooms. The place was small, with nowhere to hide. But after three hours with no results, as impossible as it might seem to not catch him, they knew he'd gotten away.

But in fact, the searchers had passed right by him, so close they might have touched him. He bided his time, waiting until they gave up, called off the search, and filed out. He smiled to himself and silently thanked the scientists who'd made it possible for him to leave the building undetected. He could do this again, if he liked, and again, and again. No one was going to find him. Once outside and away from the place, he removed his secret weapon: an invisibility shield.

That's not science fiction. Researchers in both England and the United States are already developing the mathematics and "metamaterials" that they believe will accomplish it. Metamaterials can be tuned to bend electromagnetic radiation in any direction, so a veil or "cloak" made of them that's tuned just right will not cast a shadow or reflect light. Electromagnetic radiation would simply flow around it, and people who looked at the veil would fail to see it, believing they were looking through it.

It's no surprise that the Pentagon's Defense Advanced Research Projects Agency supports this research, because there are obvious military applications, but like anything else that humans develop, sometimes it's out there before we examine the implications. Imagine what a criminal might do. Thus far, scientists haven't been able to engineer it, but the theory's in place. It may be only a matter of time before it's accomplished, especially for scientists who want it badly enough. Some predict it will occur within a year.

Hopefully, as they develop the invisibility cloak, they'll also invent a way to detect it, if needed. But then, it wouldn't be a real invisibility cloak.

The future of forensics will benefit from areas of science not yet known to be relevant but that have potential.

It takes innovative minds to see new connections, such as the vision those early scientists possessed who foresaw what could happen from their input into the legal system. To develop similar foresight today, like them, we must see beyond the obvious and anticipate new research directions, but also note potentially harmful situations.

That's what happened during the 9/11 crisis. Some 2,749 people died at the World Trade Center, Pentagon, and in Pennsylvania during the terrorist attacks on U.S. soil. Since many had been disintegrated by fire or were in small pieces scattered around the area, the task of identifying them was daunting. No agency in the world was prepared for the magnitude of such a task, and while the goal of identifying every single part of a human being was impossible, the many public and private agencies that came together tried as hard as they could to identify whomever they could. When the process was suspended in 2005, they had identified nearly 1,600. In some cases, they had to devise new techniques or new ways to process DNA more quickly.

The Bode Technology Group from Virginia, the largest DNA testing company in the country at the time, worked with the medical examiners on site. Technicians found they had to work at the time with only bone fragments, which do not yield easily to DNA analysis, and less than 40 percent of more than 12,000 fragments that went through an initial screening provided useful information for identification purposes. Bode looked into an alternative, and one process involved using decalcification to isolate the DNA, which raised the success rate. Another involved developing multiplex short tandem repeat systems, which included testing on the Amelogenin locus for gender identity. Amelogenin is the smallest of the STR markers, with about 115 base pairs. To meet the challenge, the company designed a way to make the tests more sensitive and appropriate for smaller samples.

In addition, Orchid Biosciences in New Jersey, which had done a lot of paternity testing, also made a contribution to the technology. They were using a specific type of stand-alone genetic marker and were asked to use it for identification of remains. Sometimes the only marker derived from a sample was Amelogenin, and on those samples, the company used its genetic test for genotyping information. They were able to find tissue samples from different locations around the affected area that matched one another.

Progress in forensic science is now heavily supported, and as fast as investigators can apply new discoveries in the future, that's how quick will be the advance of forensic science. As increasingly more people become educated in this field, more visionaries are likely to emerge. They may not have to fund their own projects, as some early pioneers did, but they will possess the same driving curiosity to apply technologies that can make forensic science so astonishing. Among the inventions or developments we may see in the future are the following.

Nanotechnology, the science of small things, is the up-and-coming project in science. Nanotechnologists manipulate tiny structures that would have to be expanded one thousand times just to be visible with an optical microscope. Because of their miniscule size, they can be injected into the bloodstream or absorbed through skin, and carbon nanotubes have the potential to diagnose and treat cancer, control the delivery of a drug, and even create biomechanical computers. While not directly forensic in its application, the possibility exists that this technology could enter into cases of intentional poisoning, both as a lethal substance and a means for diagnosis. Nanoparticles, the size of millionths of a millimeter, are already present in large numbers in the air from natural sources and from vehicle exhaust emissions, rubber, and copier toner. They're being considered in other applications, such as the manufacture of clothing, purifying contaminated ground, or becoming miniature security sensors.

Our current state of knowledge about the toxicology of nanoparticles and nanotubes is primitive, but experts suggest that nanoparticles may have negative effects at their point of entry into a person's body. Scientists have also observed how they travel to the central nervous system and ganglia by traveling along the axons and dendrites of neurons. Thus, nanotoxicology, which evaluates nanostructures and devices, is a developing discipline. If drugs can be delivered via nanotubes, considering the history of homicide and its intimate association with poison, it's clear that what scientists herald as an amazing new technology could become lethal—and difficult to detect.

To go along with this technology, a revised version of the atomic-force microscope, used for minute-scale measurements, is now one hundred times faster than its previous model. That could make it possible with these membrane-based probes to observe complex molecular interactions.

On a larger scale, but just as exciting, virtual imaging can produce a 3-D image of an object, which means pathologists can use it for certain autopsies without having to cut into a body. To get an overview of it for detecting damage of organs and muscles, they can employ a combination of CT scans and magnetic resonance imagining (MRI). In some places, it's already being done and as its benefits become better known, and it's viewed as more practical and cost-effective, it will likely be made available to busy morgues and hospitals. In addition, the autopsy photos will be less gruesome for juries, and there's little chance of inadvertently destroying forensic evidence inside the tissues, because there's no cutting. In addition, the digitized images can easily be stored or sent to other pathologists for consultations. It's a welcome innovation.

Japan is the leader in biometric identification. They're using palm-vein recognition technology in banks. The machines shoot a beam of light through a customer's hand, and the blood absorbs it, casting shadows that can be mapped. Customers are then filed via their personal vein map, and to remove money from their accounts, they have to allow their palms to be scanned. This technology could help stem identity theft and fraudulent bank transactions, as well as have other applications not yet envisioned.

In the area of deception detection, psychiatrist Lawrence Farwell is already using "brain fingerprinting," and claiming that it is 99.9 percent accurate. Since the brain records all human experiences, it will recognize that which it already has processed, and Farwell's device supposedly measures the presence or absence of that knowledge, including a crime scene. Similar to how a polygraph is used, the operator monitors a suspect's electrical activity via a headband equipped with sensors while the person is exposed to "prod" words or images, both relevant and irrelevant to the crime. If his resulting brainprint shows that he (or she) recognizes the relevant stimuli—a spike called a MERMER (memory and encoding related multifaceted electroencephalographic response) indicates that he has a stored memory. The absence of such a spike indicates that the person was never there. A flaw, supposedly taken care of with proper wording and a substantial amount of material about the supposed behavior in the crime, is that if a person had been at the scene but had not committed the crime, his brain might show a spike, thereby falsely implicating him. Like the polygraph, the threat of a machine that's able to detect deception has triggered confessions. As a science, it remains controversial, but will probably improve. Many parties, from the military to research psychologists to attorneys, are watching the progress of this device.

In another realm, Logicube, Inc. in California has produced CELLDEK, an information extraction device to use on cell phones and PDAs. Reportedly, it can access data from 90 percent of all North American devices. Thus, police can process cell phones and PDAs at a scene rather than sending them to labs to await the data extraction there. CELLDEK does not alter the data, and it offers information on calls made and received, the time and date, the internal phonebook, lists and memos, and even deleted material.

Among other developments on the horizon are such ideas as enhanced computer tomography for getting images from the skulls of John and Jane Does. In 1999, for example, when the dismembered remains of a woman washed onto a river's bank in Wisconsin, the skin had been peeled off her fingers and face to prevent identification. The police were loathe to subject the skull to the typical methods of forensic art, for fear of destroying evidence, so they asked the Milwaukee School of Engineering's Rapid Protoyping Center to try a technique that would yield a three-dimensional model. It took about thirty hours, but from CT scans of the head, they were able to make a computer replica and then a sculpture from thousands of thin layers of paper. This way, they could accurately render the facial features without removing soft tissue. Once that was achieved, a forensic artist took it from there and applied the usual techniques. The resulting image got an identification, which led to an arrest and conviction. In 2000, this process had never been done, but once someone in law enforcement considered how it might be, it was just a matter of engaging the scientists to apply the technology in this new way.

Speaking of computers, data carving is the process of extracting a collection of data from a larger data set. Digital investigations generally involve this procedure when experts analyze the unallocated file system space. However, the results of the existing technology often produce false positives, and thus an investigator must test each of the extracted files by opening them in an application. Computer experts wish to be able to design and develop file-carving algorithms that identify more files with better time-saving procedures and a reduced number of false positives—preferably none. In fact, one computer convention ran a contest to challenge attendees to come up with such a method. More such "games" may generate new ideas in many different areas.

Many scientific discoveries over the course of history, when applied to criminal investigation, have dramatically shifted how crimes get solved and prosecuted. There's good reason to believe we may yet see plenty of dramatic applications in the future, so we should encourage both scientific groups and law enforcement groups to ponder possibilities. But we must also take care to watch for the harm that new ideas may yield. That means being ethically accountable.

ETHICS AND SELF-POLICING

In South Wales, Britain, in 1988, a twenty-year-old woman was fatally stabbed, and with no leads, the case went cold. In 2000, with new technology in mind, detectives decided to go over her apartment once more to look for minute items of evidence that might have escaped them the first time. They were pleased to locate several tiny droplets of blood on which they could use a DNA test, and did. By this time, Britain had a national DNA data bank, so they ran the samples. While they had no cold hits, they did find a DNA profile that was a close match—for a fourteen-year-old boy who was not yet even born at the time of the murder.

However, he had relatives who'd been alive then, so the police ran DNA tests on some of them and found the match they were looking for: Jeffrey Gafoor, on the boy's father's side. Gafoor confessed, and the case was covered in the news as a "kinship analysis." Given its success, law enforcement believes that this approach will be utilized more often, especially in light of resistance to requiring the entire population to provide samples for a universal data bank. Considering studies that indicate that the chances of becoming a criminal rise when a close relative has committed a crime, the idea that DNA profiles might point down a specific path not yet recorded may look like a real boon for law enforcement.

Yet it also raises the red flag of potential civil liberties violations, since a person's privacy can be invaded through a mere association with someone whose DNA is available to the police. In Britain almost any offense allows police to collect samples for the data bank, and the pressure is growing in the United States to expand the DNA data banks under similar conditions. In addition, there are fears that certain populations, notably blacks, Hispanics, and the economically deprived, may take the greatest hit and thus be the most vulnerable.

A universal data bank would, of course, alleviate the issue of genetic surveillance, and law enforcement is generally in support of it, but most private citizens oppose it. There's a fine balance between security for the masses and the individual right to privacy, and the idea that one's essence can so easily be accessed can be daunting, especially in light of possible mishandling or misinterpretation. On the other hand, those who have been exonerated via DNA appreciate its power—and would probably welcome a data bank that would have freed them earlier, if not eliminated them as suspects altogether.

As mentioned earlier, officials at the New York–based Innocence Project, founded by Barry Scheck and Peter Neufeld, rely on DNA evidence testing to help the innocent prove their claims. They expect to have many more cases. Since there have been more than two hundred exonerations thus far, the use of this technology brings more accountability into evidence handling. They evaluate requests from prisoners or families of prisoners, read through court transcripts and other reports, and determine whether biological evidence from the relevant case has been preserved well enough for a DNA extraction. If the case fits all their criteria, and they see problems with the investigation or prosecution, they take it on. Thanks to their work, other lawyers have followed a similar pattern and many more such projects have been set up around the country to ensure a greater reach for unfortunate individuals.

For example, Dennis Williams had served eighteen years for the abduction and murder of a young couple in Illinois, placed at the scene by an eyewitness who later recanted her testimony and then reverted to her original story. He and three other men were convicted and imprisoned, two of them (including Williams) going right to death row. They might have been executed if not for a group of journalism students at Northwestern who took up the "Ford Heights Four" case in 1996. They located a witness who had reported the identity of the real killers shortly after the crime, but the police had failed to follow up on the tip. The investigating team also found two of the three men actually responsible for the crime, and they eventually confessed. DNA testing corroborated their confessions and exonerated Williams.

More than three-quarters of the falsely imprisoned were convicted in part with eyewitness misidentification, which can occur even under the best circumstances. Luis Diaz, sixty-seven, served twenty-six years for crimes he did not commit. In August 1979, Miami-Dade County police arrested him for a series of rapes over the previous two years attributed to the "Bird Road Rapist." One victim, who identified him as the killer even though he looked dramatically different from her original description, had given police his license plate number after she spotted him at a gas station. Seven other victims looked at a photo spread that featured him and agreed, despite similar inconsistencies among their descriptions (two later recanted). Diaz was convicted of seven rapes and attempted rapes. Only two semen samples were recovered, but that was sufficient for later DNA analysis. Diaz was cleared. The need for innocence watchdogs is apparent.

On October 30, 2004, President George W. Bush signed the "Justice for All" act, which set up guidelines and enhanced funding for the use of DNA technology in the legal arena. Convicted felons gained the right to post-conviction testing if DNA evidence is available and provides a reasonable chance of proving the person did not commit the crime. States and other localities will receive money to do more such testing and improve their labs. In addition, the labs must undergo accreditation every two years to comply with guidelines, and CODIS will be expanded to allow state crime labs to send lawfully collected samples to the data bank, including from juveniles adjudicated delinquent.

And speaking of CODIS, the FBI laboratory has partnered with four regional crime labs in Arizona, Minnesota, Connecticut, and New Jersey to increase the capacity to test for mtDNA, which is inherited through the matrilineal line. This effort will assist in identifying John Does and matching them with missing persons, and perhaps assist with identification in terrorism cases. Prior to this development, only the FBI lab performed mtDNA examinations free of charge. Partnering with these labs and paying expenses will double the FBI's capabilities.

Among other concerns discussed at professional conferences and in forensic science chat rooms is the need for higher standards, forums of accountability that will be taken seriously, and a filter against junk science that harms the innocent and assists the guilty. There's also a need for better research about how and when courtroom procedures may undermine the cause of justice.

For example, "earprint testimony" put a British man in prison for life. He had served seven years before the "scientific evidence" was exposed as flawed. Mark Dallagher, thirty-one, was convicted at Leeds Crown Court in 1998 of the murder of ninety-four-year-old Dorothy Wood. The prosecution expert, Cornelis Van Der Lugt, told the court he was "absolutely convinced" that the earprints found on window glass where the perpetrator entered the victim's home were made by Dallagher's ears. The case made legal history as the first one in which earprints led to a successful prosecution, and one prosecutor described it as "a great step forward for forensic science." But then a DNA profile obtained from the earprint proved that it was not Dallagher's after all. Thus, instead of a step forward, it proved to be an embarrassing step backward. Indeed, as judges take on the gatekeeper's duties for deciding when expert testimony is genuinely science or not, they must get more educated themselves in the sciences. Studies indicate that this has not been the case, and the admissibility of "science" in some places has been rather haphazard. For the most part, they fail to understand the notion of hypothesis testing and error rates, so they often ignore these criteria.

Along these same lines, numerous crime labs around the country have been closed or notified that they must clean up their act, after examination of cases revealed poor controls, mishandling of evidence, and clumsy mix-ups. Worse, a few labs employed people who failed to do the tests they claimed they had, or who fabricated results.

Between the need to keep judges educated about actual science for admissibility and concern over crime lab protocol, forensic scientists have called for greater accountability and the need for better education. With so many people proven innocent whose lives have been decimated by the legal process, we need to be as careful about our forensic projects as we are enthused about innovations on the horizon. Since terrorist attacks and the world's destabilization have shown us there's much more at stake these days than was once the case, and forensic science will probably be a part of any major world catastrophe, we need to develop ethical applications from good, solid science, for use on a global scale.

GLOBALIZATION

Besides innovation, forensic scientists are seeking ways to expand their applications to more countries, especially those in need that have few or no resources. The FBI assists internationally with setting up computerized databases and teaching their programs, while specially trained teams arrive after a natural disaster or massacre to assist with the identification of the dead. With biological terrorism always looming, those with knowledge of microbial technology can assist in developing devices for tracking and for protection. In the field of informatics, we can develop more accurate threat assessment programs, as well as prepare ourselves to respond to large-scale events, as we did with the September 11 incidents.

As part of that effort, Electronic Sensor Technology in California has developed the zNose®, a gas chromatograph that captures and analyzes odors, based on "Surface

Acoustic Wave (SAW)" technology. One model is portable, and three other types are in development. This device will work well for homeland security, as units are installed in buildings as early warning alarms for chemical or biohazardous threats. Other smell recorders are also in the works.

In addition, there's a global alert map, operated by the National Association of Radio-Distress Signaling and Infocommunications, in Hungary, that provides an almost real-time recap of events as they occur. When a man takes over a one-room schoolhouse in the Amish country in Pennsylvania, the entire world can watch what happens. This also means that forensic scientists with certain specialties can be alerted almost immediately if they're needed in some area far from them.

We've come a long way from convictions based on "spectral evidence," criminal identification with torture, and the detection of lies via chewing rice. What happens from here might only be limited by the boundaries of our imaginations. While it may be "better that ten guilty persons escape than that one innocent suffer," as English jurist William Blackstone stated, it would be better yet if we could avoid both. With biology, psychology, odontology, entomology, and the other areas of forensic science all pulling together, we may one day achieve this ideal. We have our forensic pioneers to thank for that.

SELECTED BIBLIOGRAPHY

Allen, Matthew. *Essay on the Classification of the Insane.* London: John Taylor, 1837.

Baden, Michael, with Marion Roach. *Dead Reckoning: The New Science of Catching Killers.* New York: Simon & Schuster, 2001.

———with Judith Adler Hennessee. *Unnatural Death: Confessions of a Medical Examiner.* New York: Ivy Books, 1989.

Barrett, Sylvia. *The Arsenic Milkshake.* Toronto, Ontario, Canada: Doubleday, 1994.

Bass, Bill, with Jon Jefferson. *Death's Acre: Inside the Legendary Forensic Lab Where the Dead do Tell Tales.* New York: G. P. Putnam's Sons, 2003.

Beavan, Colin. Fingerprints. New York: Hyperion, 2001.

Becker, Peter and Richard F. Wetzell. *Criminals and Their Scientists: The History of Criminology in International Perspective.* Cambridge, England: Cambridge University Press, 2006.

Behn, Noel. *Lindbergh: The Crime.* New York: NAL-Dutton, 1994.

Berg, A. Scott. *Lindbergh.* New York: G. P. Putnam's Sons, 1998.

Block, E. *The Wizard of Berkeley.* New York: Coward-McCann, 1958.

Borowitz, Albert. *Blood and Ink: An International Guide to Fact-Based Crime Literature.* Kent, OH: Kent State University Press, 2002.

Botz, Corinne M. *The Nutshell Studies of Unexplained Deaths.* New York: Monacelli Press, 2004.

Boucher, Anthony. *The Quality of Murder.* New York: E. P. Dutton & Co., Inc., 1962.

Britz, Marjie T. *Computer Forensics and Cyber Crime.* Upper Saddle River, NJ: Prentice-Hall, 2004.

Brown, Douglas and E. V. Tullett. *The Scalpel of Scotland Yard: The Life of Sir Bernard Spilsbury.* New York: E. P. Dutton and Company, Inc., 1952.

Bullock, Alan, Ed. *World History: Civilization from its Beginning.* Garden City, New York: Doubleday & Company, Inc., 1962.

Carlson, Oliver. *James Gordon Bennett: The Man Who Made the News.* New York: Duell, Sloan, and Pearce, 1942.

Carr, John Dickinson. *The Life of Sir Arthur Conan Doyle.* New York: Carroll & Graf Publishers, 1949.

Cleckley, H. *The Mask of Sanity*, 5th edition. St. Louis, MO: Mosby, 1976.

Colaizzi, Janet. *Homicidal Insanity, 1800–1985.* Tuscaloosa, AL: University of Alabama Press, 1989.

Cole, Simon A. *Suspect Identities: A History of Fingerprinting and Criminal Identification.* Cambridge, MA: Harvard University Press, 2001.

Cornwall, John. *Hitler's Scientists: Science, War and the Devil's Pact.* New York: Penguin, 2003.

Cunningham, Aimee. "Particular Problems: Assessing the Risk of Nanotechnology," *Science News*, Vol. 169. No. 18. May 6, 2006.

Davies, Norman. *Europe: A History.* New York: Harper, 1996.

DeNevi, Don and John H. Campbell. *Into the Minds of Madmen: How the FBI Behavioral Science Unit Revolutionized Crime Investigation.* Amherst, NY: Prometheus Books, 2004.

Diagnostic and Statistical Manual of Mental Disorders—IV. Washington, D.C.: American Psychiatric Association, 1994.

De River, J. Paul. *The Sexual Criminal.* Springfield, IL: Charles C. Thomas, 1949.

Edwards, Samuel. *The Vidocq Dossier.* Boston, MA: Houghton Mifflin Company, 1977.

Evans, Collin. *The Casebook of Forensic Detection.* New York: John Wiley & Sons, Inc., 1996.

———. *A Question of Great Forensic Controversies, from Napoleon to O. J.* Hoboken, NJ: John Wiley & Sons, Inc., 2003.

———. *The Father of Forensics: The Groundbreaking Cases of Bernard Spilsbury and the Beginnings of Modern CSI.* New York: Penguin, 2006.

———. *The Second Casebook of Forensic Detection.* Hoboken, NJ: John Wiley & Sons, Inc., 2004.

Everitt, David. *Human Monsters.* Chicago: Contemporary Books, 1993.

Faigman, David L. *Legal Alchemy: The Use and Misuse of Science in the Law.* New York: W.H. Freeman and Company, 1999, 2000.

Fido, Martin. *The Chronicle of Crime.* London: Carlton Books, 1999.

Field, Kenneth S. *History of the American Academy of Forensic Sciences, 1948–1998.* Conshohocken, PA: American Society for Testing and Materials, 1996.

Fisher, Jim. *The Lindbergh Case.* Princeton, NJ: Rutgers University Press, 1994.

Frasier, David K. *Murder Cases of the Twentieth Century.* Jefferson, NC: McFarland & Company, 1996.

Fridell, Ron. *Solving Crimes: Pioneers of Forensic Science.* New York: Grolier, 2000.

Friedman, Lawrence M. *Crime and Punishment in American History.* New York: Basic Books, 1993.

Futrelle, Jacques. *The Thinking Machine.* New York: Random House, 2003.

Gerber, Samual M. and Richard Safterstein, Eds. *More Chemistry and Crime.* Washington, D.C.: American Chemical Society, 1997.

Goff, M. Lee. *A Fly for the Prosecution: How Insect Evidence Helps Solve Crimes.* Cambridge, MA: Harvard University Press, 2000.

Golan, Tal. *Laws of Men and Laws of Nature.* Cambridge, MA: Harvard University Press, 2004.

Gollmar, Robert. *Edward Gein: America's Most Bizarre Murderer.* New York: Pinnacle, 1981.

Gribbin, John. *The Scientists.* New York: Random House, 2002.

Grun, Bernard. *The Timetables of History, Third Edition.* New York: Simon & Schuster, 1991.

Hakim, Joy. *The Story of Science: Aristotle Leads the Way.* Washington, D.C.: Smithsonian Books, 2004.

Halberstam, David. *The Fifties.* New York: Villard, 1993.

Hare, R. D. *The Psychopathy Checklist--Revised, 2nd Edition.* Toronto, Ontario, Canada: Multi-Health Systems, 2003.

Helmer, William with Rich Mattox. *Public Enemies: America's Criminal Past.* New York: Checkmark, 1998.

Horn, David G. *The Criminal Body: Lombroso and the Anatomy of Deviance.* New York: Routledge, 2003.

Houck, Max M. *Trace Evidence Analysis.* New York: Elsevier, 2004.

Hughes, Rupert. *The Complete Detective.* New York: Sheridan House, 1950.

Inman, Keith, and Norah Rudin. *An Introduction to Forensic DNA Analysis.* Boca Raton, FL: CRC Press, 1997.

Innes, Brian. *Bodies of Evidence.* Pleasantville, NY: Reader's Digest Press, 2000.

———. *Profile of a Criminal Mind.* Pleasantville, NY: Reader's Digest Press, 2003.

James, Stuart H. and Jon J. Nordby, *Forensic Science: An Introduction to Scientific and Investigative Techniques.* Boca Raton, FL: CRC Press, 2003.

Jones, David A. *History of Criminology: A Philosophical Perspective.* Westport, CT: Greenwood Press, 1986.

Kennedy, Lodovic H. *Crime of the Century.* New York: Viking, 1996.

———. *Ten Rillington Place.* New York: Simon & Schuster, 1961.

Krafft-Ebing, Richard von. *Psychopathia Sexualis with Especial Reference to the Antipathic Sexual Instinct: A Medico-Forensic, Revised Edition.* Philadelphia: Physicians and Surgeons, 1928.

Lane, Brian. *The Encyclopedia of Forensic Science.* London: Magpie Books, 1994, 2004.

Larson, Erik. *The Devil in the White City.* New York: Crown, 2003.

Lee, Henry C. and Frank Tirnady. *Blood Evidence: How DNA Revolutionized the Way We Solve Crimes.* Cambridge, MA: Perseus, 2003.

Levy, Harlan. *And the Blood Cried Out.* New York: Basic Books, 1996.

MacDonell, Herbert L. *Bloodstain Patterns.* Corning, NY: Laboratory of Forensic Science, 1993.

Masters, R. E. L. and Eduard Lea. *Perverse Crimes in History.* New York: The Julian Press, 1963.

Miller, Hugh. *Proclaimed in Blood: True Crimes Solved by Forensic Scientists.* London: Headline, 1995.

———. *What the Corpse Revealed: Murder and the Science of Forensic Detection.* New York: St. Martin's Press, 1998.

Mohr, James C. *Doctors and the Law.* Baltimore, MD: The Johns Hopkins University Press, 1993.

Morland, Nigel. *An Outline of Scientific Criminology.* Second Completely Revised Edition, New York: St. Martin's Press, 1971.

Morse, Andrew. "New Biometric Identifier Is At Hand." *The Wall Street Journal.* July 21, 2005.

Morton, James. *Catching the Killers: A History of Crime Detection.* London: Random House, 2001.

Nickell, Joe and John Fischer. *Crime Science: Methods of Forensic Detection.* Lexington, KY: The University Press of Kentucky, 1999.

Nordby, Jon. *Dead Reckoning: The Art of Forensic Detection.* Boca Raton, FL: CRC Press, 2000.

Owen, David. *Hidden Evidence: Forty True Crimes and How Forensic Science Helped Solve Them.* Buffalo, NY: Firefly Books, 2000.

Parry, Leonard. *Some Famous Medical Trials.* New York: Charles Scribner's Sons, 1928.

Perry, Marvin. *A History of the World.* Lawrenceville, NJ: Houghton Mifflin Co., 1985.

Platt, Richard. *The Ultimate Guide to Forensic Science.* London: DK Publishing, 2003.

Ramsland, Katherine. *The CSI Effect.* New York: Berkley, 2006.

———. *The Forensic Science of CSI.* New York: Berkley, 2001.

———. *The Human Predator: A Historical Chronicle of Serial Murder and Forensic Investigation.* New York: Berkley, 2006.

———. *The Science of Cold Case Files.* New York: Berkley, 2001.

Rhine, Stanley. *Bone Voyage: A Journey in Forensic Anthropology.* Albuquerque, NM: University of New Mexico Press, 1998.

Roughead, William. *Classic Crimes.* New York: New York Review of Books, 2000.

Rumbelow, Donald. *Jack the Ripper: The Complete Casebook.* New York: Contemporary Books, 1988.

Saferstein, Richard. *Criminalistics: An Introduction to Forensic Science, 9th Edition.* Upper Saddle River, NJ: Prentice-Hall, 2006.

Scaduto, Anthony. *Scapegoat: The Lonesome Death of Bruno Richard Hauptmann.* New York: G. P. Putnam's Sons, 1976.

Scheck, Barry, Peter Neufeld, and Jim Dwyer. *Actual Innocence.* New York: Random House, 2000.

Shaler, Robert C. *Who They Were: Inside the World Trade Center DNA Story: The Unprecedented Effort to Identify the Missing.* New York: The Free Press, 2005.

Sifakis, Car. *The Encyclopedia of American Crime.* New York: Facts on File, 1982.

Snyder Sachs, Jessica. *Corpse: Nature, Forensics and the Struggle to Pinpoint Time of Death.* Cambridge, MA: Perseus, 2001.

Stocking, Jr., George W. *Victorian Anthropology.* New York: Free Press, 1987.

———. *Bones, Bodies, Behavior: Essays on Biological Anthropology, Volume 5.* Madison, WI: University of Wisconsin Press, 1988.

———. *The Ethnographer's Magic and Other Essays in the History of Anthropology.* Madison, WI: The University of Wisconsin Press, 1992.

Sung, Tz'u. *The Washing Away of Wrongs.* Translated by Brain E. McKnight. Ann Arbor, MI: Center for Chinese Studies, University of Michigan, 1981.

Symons, Julian. *A Pictorial History of Crime: 1840 to the Present.* New York: Bonanza Books, 1966.

Taylor, Karen T. *Forensic Art and Illustration.* Boca Raton, FL: CRC Press, 2000.

Thomas, Ronald. *Detective Fiction and the Rise of Forensic Science.* Cambridge, England: Cambridge University Press, 1999.

Thorwald, Jurgen. *The Century of the Detective*. New York: Harcourt, Brace & World, 1964.

———. *Crime and Science*. New York: Harcourt, Brace & World, 1966.

Trestrail, John Harris. *Criminal Poisoning*. Totowa, NJ: Humana Press, 2000.

Tullet, Tom. *Strictly Murder: Famous Cases of Scotland Yard's Murder Squad*. New York: St. Martin's Press, 1979.

Ubelaker, Douglas and Henry Scammel. *Bones: A Forensic Detective's Casebook*. New York: M. Evans and Company, 1992.

Vale, Gerald. "History of Bitemark Evidence." *Bitemark Evidence*, Robert J. Dorian, Ed. New York: Marcel Dekker, 2005.

Vidocq, Francois Eugéne. *Memoirs of Vidocq: Master of Crime*. Edinburgh, Scotland: AK Press, 2003.

Vitray, Laura. *The Great Lindbergh Hullabaloo: An Unorthodox Account*. New York: William Fargo, 1932.

Wambaugh. *The Blooding: The True Story of the Narborough Village Murders*. New York: William Morrow & Co., 1989.

Whipple, Sydney. *The Lindbergh Crime*. New York: Blue Ribbon Books, 1935.

Wilkes, Roger. *The Book of Murder and Violence, Vol 1*. London: Robinson, 2000.

Wilson, Colin. *Murder in the 1940s*. New York: Carroll & Graf, 1993.

Wilson, Colin and Damon Wilson. *Written in Blood: A History of Forensic Detection*. New York: Carroll & Graf, 2003.

Wonder, A.Y. *Blood Dynamics*. San Diego, CA: Academic Press, 2001.

INDEX